ABOUT ISLAND PRESS

ISLAND PRESS, a nonprofit organization, publishes, markets, and distributes the most advanced thinking on the conservation of our natural resources—books about soil, land, water, forests, wildlife, and hazardous and toxic wastes. These books are practical tools used by public officials, business and industry leaders, natural resource managers, and concerned citizens working to solve both local and global resou᷄ ᷄ problems.

Founded in 1978, Island Press reorganized in 1984 to ι.   'he increasing demand for substantive books on all resource-related ιϩ.   'and Press publishes and distributes under its own imprint and ofte.   ιϲϩ services to other nonprofit organizations.

Funding to support Island Press is provided by The Mary Reynolds Babcock Foundation, The Ford Foundation, The George Gund Foundation, The William and Flora Hewlett Foundation, The Joyce Foundation, The J. M. Kaplan Fund, The John D. and Catherine T. MacArthur Foundation, The Andrew W. Mellon Foundation, Northwest Area Foundation, The Jessie Smith Noyes Foundation, The J. N. Pew, Jr. Charitable Trust, The Rockefeller Brothers Fund, The Florence and John Schumann Foundation, and The Tides Foundation.

Nowhere else on Earth does a continental,
temperate-climate fauna naturally exist stranded
in a tropical, oceanic-island flora.

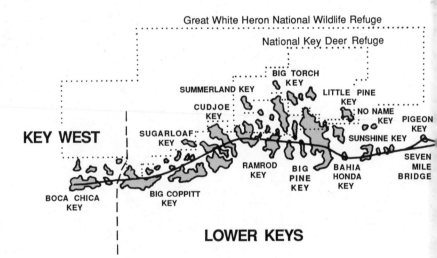

WILDLIFE
of the FLORIDA
KEYS: A NATURAL
HISTORY

MIAMI

KEY
LARGO

Gulf of
Mexico

TAVERNIER

PLANTATION KEY

UPPER
MATECUMBE

WINDLEY KEY

MARATHON

FIESTA
KEY

LAYTON

GRASSY   CONCH
KEY   KEYS

VACA
KEY

LOWER
MATECUMBE

LONG KEY

DUCK KEY

CRAWL
KEY   KEY
KEY
COLONY
BEACH

ISLAMORADA

Atlantic
Ocean

*While sailing towards the Florida Keys, my mind
was agitated with anticipation of the delight I should
experience in exploring a region whose productions
were very imperfectly known.*

JOHN JAMES AUDUBON, 1832

WILDLIFE of the FLORIDA KEYS: A NATURAL HISTORY

by
James D. Lazell, Jr.

ISLAND PRESS
Washington, D.C. □ Covelo, California

Cover photographs by Bill Keogh (clockwise from left) Key deer, Dama clavia; Cuban tree frog, Hyla septentrionalis; Tricolored heron, Egretta tricolor.
Cover design by Sue Rose
Text design by Irving Perkins Associates

Library of Congress Cataloging-in-Publication Data

Lazell, James D.
  Wildlife of the Florida Keys.

  Includes index.
  1. Island fauna—Florida—Florida Keys—Ecology.
  2. Island fauna—Florida—Florida Keys—Identification.
  3. Nature conservation—Florida—Florida Keys. I. Title.
  QL169.L39 1989   591.9759'41   89-1780
  ISBN 0-933280-98-X
  ISBN 0-933280-97-1 (pbk.)

Printed on recycled, acid-free paper

Manufactured in the United States of America

10 9 8 7 6 5 4 3 2 1

Dedicated to the memory of seven people who as outdoorsmen, naturalists, or scholars profoundly influenced my life:

JOSEPH M. CADBURY
ARCHIE CARR
JOHN F. MONTGOMERY
DICK D. QUIN
GEORGE GAYLORD SIMPSON
RICK WARNER
and, of course,
JACK WATSON

ACKNOWLEDGMENTS

This work has been made possible through a grant from The Conservation Agency. All royalties from the sale of this book go directly to that Agency to support scientific research in wildlife conservation. The Conservation Agency pays no salaries. Some expenses have been defrayed, but all of the photography and artwork herein was donated to the cause of making facts about Florida Keys wildlife accessible to the public.

So many people have contributed to this work that I feel sure I will leave someone out in the list that follows; I apologize in advance. The contributors fall into broad categories, not designated individually. Those who gave photographs or original art are acknowledged here and by their works. Others read, edited, updated, and revised major sections (such as the bird list) for me. Many worked with me in the field, actually catching animals, preparing specimens, or transporting me to remote Keys. Some curated the material once it got back to the Museum. A few—perhaps the most noble—save specimens. They do it in my absence, in the firm belief that next month or next year I will return. They usually do it by donating part of their household freezer space to a road-killed rabbit, rat, or rattlesnake. There are not many of them, but I depend on them. In the end science depends totally on the tangible—what one can count and measure. Without those specimens it is all hearsay and speculation.

So, to all of you my gratitude and my hope that our efforts will aid the wildlife of these little islands we love: Ross Arnett, Peter Auger, the late Stuart Avery, Joel Beardsley, Bonnie Bell, Sheryl Ives Boynton, Marge and L. Page Brown, Walt Carson, Steve Christman, Jonathan Coddington, John Crawford, Dawn D'Alessandro, Liz Dann, Ken Dodd, William Dunson, Robert Fisher, Bill and Fran Ford, Carol Fries, George Garrett, Tricia Giovannone, Jeff Goodyear, Numi Spitzer Goodyear, Jim Hardiman, Bob Head, Ralph Heath, Kathy Hill, Andy Hooten, Ralph Jess, Artis Johnston, Lois Kitching, Jody and Henry Klaassens, Karl Koopman, Tom Kruger, Larry Lawrence, James Layne, Laura Leibensperger, Scott Miller, Robert Mochilar, Paul Moler, Carol Munder, Nancy Nielsen, Joe Oliver, Oscar Owre, Jeanne Parks, Brandy Pontin, Louis Porras, George Powell, Robert Rackley, Mark Robertson, William Robertson, Pat Rogers, Jose Rosado, Franklin Ross, Dave Ryan, Jeff Schaeffer, Albert Schwartz, Fred Sibley,

Charlotte Smith, Jan Soderquist, Peter Warny, the late Thurlow Weed, Art Weiner, Susan Whiting, and Roger Wood.

Three other organizations have helped fund our research: Earthwatch, The Explorers Club, and The Nature Conservancy.

Three of my peers have reviewed and critically analyzed this book in manuscript: Ross Arnett, William Robertson, and Arthur Weiner. They know more about my subject than any other people I know. I have not incorporated every one of their suggestions in my several revisions, but I have included the vast majority. Of course I am still responsible for my opinions and mistakes.

	James D. Lazell, Jr.
The Conservation Agency	Snake Acres
6 Swinburne Street	Middle Torch Key
Jamestown	Monroe County, Florida
Rhode Island 02835	

CONTENTS

PREFACE

"George was right. The Keys ARE different than anything I've ever seen."

HUMPHREY BOGART, IN *KEY LARGO*, 1948

Only a few gifted individuals truly grasp the uniqueness of the Florida Keys as quickly as Bogart's character did in Maxwell Anderson's post-war melodrama. For most tourists, exposure to the natural and cultural history of the Florida Keys is a jet ski ride past a crowded beach. Snowbirds, year-round residents, and even the native "conchs" also seldom recognize the significance of the tremendous biological diversity contained in the shallow waters and coral islands of this tropical archipelago.

But who is to blame? The Florida Keys cover over one thousand square miles—whether you're a scientist or self-taught wildlife enthusiast, that's a lot of territory to become familiar with. Also, biogeographic-related taxonomic complexities have caused many researchers and learned naturalists to either lump the Keys' wildlife "with the rest of Florida" or move on to a more straightforward assignment. Until now, those interested in the wildlife of the Keys had to carry around one-half of a well-stocked library to reference the islands' fauna.

It often takes someone special to accomplish something special—to cause a leap forward in the understanding of a complex situation. In this case the someone special is Dr. James Draper Lazell, Jr. and the something special is this book, *Wildlife of the Florida Keys: A Natural History*.

The study of the wildlife of the Florida Keys was custom-made for "Skip" Lazell. He is a field biologist, ecologist, herpetologist, mammologist, and island biogeographer. But perhaps most of all he is an explorer—an explorer of not only wild places but of intellectual wildernesses as well. In this book he has brought all of his talents and skills together to vividly examine the complex, unique, and beautiful array of life forms that has evolved in this environment. Contained herein are descriptions, life histories, and anecdotes that tell a fascinating yet often troubling story. Not only will readers of this book come away with a sense of the animals that live here, but also with a sense of responsibility for the future of these fellow island dwellers.

Over the past 400 years, written observations of the wildlife of the Florida Keys have reflected the diversity of those recording the information and the circumstances under which they worked. Do. d'Escalante Fontaneda, a captive survivor from a (*circa*) 1557 Spanish shipwreck, provided fascinating accounts of the deer, bear, raccoons and other Keys fauna from a utilitarian perspective unique to his Caloosa captors. In 1799, Andrew Ellicott, a U.S. commissioner, spent six days on Key Vaca (Marathon) and recorded the killing of several deer "of that small species common to some of these islands . . ." (Key Deer) to feed his starving troops.

This book, too, is about survival. But unlike all previous works, this one voices a strong concern for the survival of the wildlife of the Florida Keys.

One wonders if there is still time.

Chuck Olson
Executive Director
Florida Keys Land Trust
Big Pine Key

WILDLIFE
of the FLORIDA
KEYS: A NATURAL
HISTORY

INTRODUCTION

The Florida Keys should have been a National Park.

THE MIAMI HERALD, CA 1970

Trailing away from Cape Florida for more than 200 miles is a vast arc of low islands. They are America's entrée to the sunny Caribbean: the only West Indies one can drive to.

Before the turn of this century, the West Indies became focal in the studies of American biologists. As theaters of evolution, these complex archipelagos far exceeded Darwin's Galapagos in richness and diversity. They nearly rivaled Wallace's Malay Archipelago, but were far closer to home and generally more accessible. Much of the matter of modern evolutionary theory, population biology, and theoretical ecology was mined from limey lowlands and grand craggy peaks of the West Indies.

That vast arc of low islands—the Florida Keys—received relatively little attention compared to sister archipelagos such as the Bahamas, Cuba, and Hispaniola and their multitude of satellite isles, and even the remote Lesser Antilles.

There were auspicious starts. Men from such places as clammy basements and gloomy attics of Cambridge, Massachusetts, came south or established contacts and sought specimens. In 1880 Samuel Garman described and named a bizarre new sea turtle for the man who sent him the first specimen. It was Kemp's ridley, named for Richard Kemp, a Bahamian who had settled in Key West. The indefatigable Outram Bangs established himself on a South Florida east coast barrier island and made forays into the subtropical interior. Among his many additions to the lists were the then-new Florida raccoon and—most spectacularly—the magnificent Florida panther. In 1899 he named the latter for the great hunter-naturalist Charles B. Cory. In 1920 Thomas Barbour, a herpetologist with much wider enthusiasms, reached the Lower Keys. He managed to bag two Key deer, which in the paper he wrote with Harvard's great mammalogist Glover Allen became the basis of the description of this novel "toy" form.

The pattern in the Keys was similar to that elsewhere in the West Indies: men from the north came, took specimens, and retreated. Local interest,

1

leading to real scientific research, developed slowly and sparsely. During the years after World War II—the fat years for scientific research in America—vast effort was expended on the West Indies south and east. Almost nothing happened in the Keys. I first came to South Florida in 1957, but I looked for things already known. The magnitude of the unknown never crossed my mind.

I returned, briefly, in 1972, leading a group of young students. Most of my career until then had been spent discovering and studying the fauna of the West Indies, but I did not believe there was much to be added in these Keys. Just how wrong I was became clear when one of those students, a sixteen-year-old girl, insisted on searching for small, aquatic rodents (belonging to a group widespread, but virtually extinct, in the West Indies) on a far Lower Key. She succeeded, and thereby discovered a new species: the silver rice rat.

In those really quite proximate days the Florida Keys were hardly studied scientifically, and the works of man still lay lightly on the land. The changes since then have been amazing, appalling, and exemplary.

There have been amazing scientific discoveries: the new rice rat and a new rabbit have been described and named; the wonderful tropical leaf-nosed bat family has been added to the fauna of Florida; major range extensions have been chronicled and new populations of rare animals discovered.

The hand of man has struck these little islets with ferocity and mindless greed. The native vegetation has been stripped, the land scarified. Booming populations have flowed down the new, broadened highway. No biologist seriously objects to people visiting the Keys. Most of us—now and historically—are but visitors here ourselves. But these Keys cannot possibly house all the migrants of America, all those looking for a temporary or permanent place in the sun. Growth has limits.

The Florida Keys are a perfect example of a worldwide phenomenon. As human population rises like a flood tide, developers rush into "undeveloped" areas. Seldom is their concern designing *with* nature, but more often in destroying as much of nature as they can before anyone has a chance to catalogue its diversity or document its beauty. The quick buck is perceived with short sight. In the degradation of the ecosystems that support us lies our doom, or the doom of our descendants.

My purpose in writing this book is to stimulate interest—among local people and visitors, scientists and laymen—in this remarkable fauna. I have tried to include all the mammals, reptiles, and amphibians (except species kept as captive pets), but I have probably failed because our surrounds—mainland Florida, the Bahamas, and Cuba—leak animals to the Keys. Things change and no range map or list is ever final.

I have cited bird guides that include the Keys and reviewed their relative merits. I have given accounts of special birds: those you are unlikely to see anywhere in the United States except these Keys, or which are much rarer elsewhere. There is included an annotated checklist of all known birds.

The world of invertebrates is huge beyond anyone's knowledge. Most of Earth's insects, for example, have yet to be discovered scientifically and even named. I have included large, common species, ones that incite fear or horror in many people, and some rare, beautiful species too. This book is about *terrestrial* life: life of the land. My limit of coverage is set by the edge of the sea: where the sea begins the Keys end. An excellent accompanying reference, therefore, is L. Gilbert Voss, *Seashore Life of Florida and the Caribbean*. 2d ed. (Miami: Banyan Books, 1980).

To illustrate the polemics that divide the scientific community with respect to the Florida Keys, consider my 1987 publication, "Conservation of a Florida habitat," *Conservation Biology* 1(3):268–69. This review of the North Key Largo Habitat Conservation Plan questions the motives of the planners and the scientific basis of their recommendations, which call for greater density of development than existing law allows. Compare S. Humphrey's 1988 publication, "The Habitat Conservation Plan," *Conservation Biology* 2(3):240–44, a defense of the same plan I oppose. To find out who is right will require independent analysis of the data, the plan as written, and the law.

THE LAND

To this day . . . there remains in geology plenty of room for the creative imagination.

JOHN MCPHEE (1983)

Controversy haunts and plagues the Florida Keys. Controversy here begins at the very beginning, with historical geology. The Keys, like the Bahamas, are made of oceanic limestone. The deeper limestones may be millions of years old, but the surficial material is very young—geologically speaking. That all this limestone was deposited on solid, igneous, continental-type strata is a sore point for the disciples of continental drift. The basement rock of the Bahamas and Florida gets right in the way of Africa, if you believe in primordial Pangaea. A half dozen disparate plate tectonic schemes have been promulgated for the Caribbean Basin. They contain mutually exclusive factors of time and motion, position and distance; none can account for Florida and the Bahamas.

In any case, from a terrestrial faunal standpoint, the geological history of the Florida Keys cannot go farther back than one hundred thousand years. At that time, during the Sangamon Interglacial, before the last great Ice Age, all of our area of interest was under water. Because little of Earth's water was frozen in the form of glacial ice, sea level was about twenty-one meters (or 70 feet) above its present position. Florida was reduced to a few islands; the closest was away to the north of present-day Lake Okeechobee, where Highlands County is today.

Great coral reefs flanked the submerged southern Florida plateau, as lesser ones do today. Behind these reefs, in the rather still, warm waters on the Gulf side, countless tons of lime precipitated from decomposing calcareous algae and other sources. These sediments of lime particles were rolled back and forth by tidal currents and eventually became rounded. The rounded lime pellets, seen under a microscope, resemble tiny eggs. Therefore they make up what is referred to in the "latineze" of scientists as egg stone: *oolite*. I was taught to pronounce each "o" separately and long, with the accent on the first: o-o-lite—making three syllables. In South Florida most folks pronounce the "o's" together, as in "too"—making a two-syllable word. It is not a moral issue.

The weight of the sediments of lime pellets, steadily increasing beneath the sea, compacted the material into concrete-like rock: oolitic limestone, or simply oolite.

No one knows for sure just what triggers an Ice Age. Beginning something like eighty thousand years ago, more ice began to accumulate around the poles than melted each year. Ice began to spread out and flow—literally if slowly—into previously temperate latitudes. The great Wurm (or "Wisconsinan") Glaciation began. As the growing ice caps consumed more and more of Earth's water, sea level dropped. The seaward-flanking reefs and oolite-capped plateau broke the falling surface. First islands, then a vast plain, appeared, open to colonization by land plants and animals.

Those islands in the falling Sangamonian sea were the primordial Florida Keys. They were surely colonized over water, in the usual manner of the major West Indian island clusters, or banks. Plants, insects, a few reptiles, probably a mammal or two (like rice rat and fruit bat), possibly even some amphibians, reached the Keys as waif dispersers, across water. If sea level had remained static at a point corresponding to its present level, the fauna of the Florida Keys would be quite different—and much more like that of the low cays of the Bahamas and Antilles. It would be a much simpler fauna, and, for me, much duller. But sea level continued to drop; more and more land appeared; continuity with the Florida mainland was achieved; a veritable zoological garden wandered onto the plain studded with high points that are our Keys.

During the Wurm glacial maximum South Florida extended, as solid

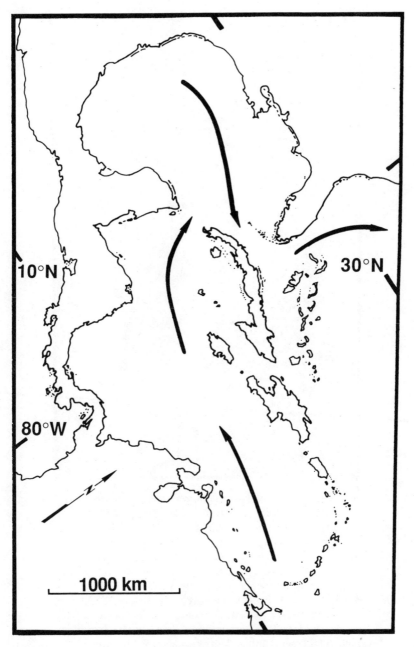

The Caribbean Basin. Arrows indicate the principal directions of surface current flow. The equatorial current enters the basin in the southeast. The Gulf Stream exits between Florida and the Bahamas in the northeast.

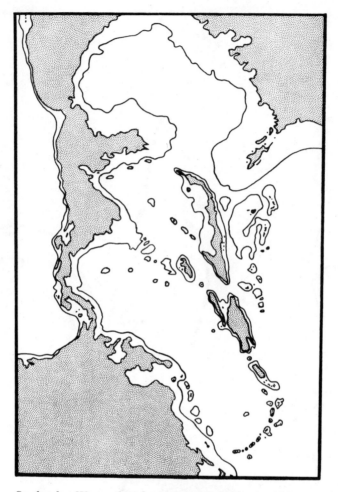

Sea level at Wurm glacial maximum (solid line) and land
areas during the Sangamon Interglacial (stippled).
Compare to the previous map of the Caribbean Basin today.
The currents remain the same, as does the quantity of the
moving water. During the Wurm the velocity of the Stream
dramatically increased. This map was difficult to draw, and
is only vaguely accurate, especially regarding the
Sangamon Interglacial level. No charts show a 21 m (70 ft)
contour. However, sea level changes affected Yucatan,
Florida, and the Bahamas far more dramatically than they
have the Lesser Antilles. Most biogeographers rarely
consider distance and land area changes in the last
100,000 years shown here.

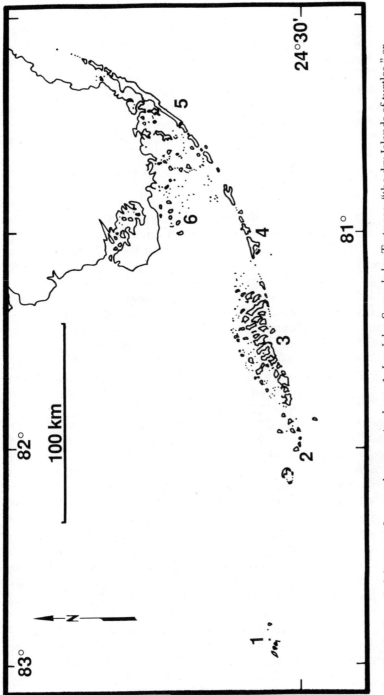

The Florida Keys. Subdivisions frequently recognized are: 1, *Las Islas Secas de las Tortugas:* "the dry Islands of turtles," or the Dry Tortugas. These are usually considered part of the next group. 2, the Outer or Sand Keys. 3, the Lower Keys, made largely of oolite. 4, the Middle Keys and 5, the Upper Keys, both made of Key Largo limestone. 6, the Bay or Mud Keys.

land, away beyond the far Tortugas to the south and west. Camels, peccaries, giant sloths, and bears roamed glades and savannahs where bone fish and barracuda prowl the turtle grass today. The sea fell 120 meters (nearly 400 feet) below its present level. Most of Canada and New York, and all of New England, were buried beneath glacial ice. Rivers of melt water poured south from the ice, flowed hundreds of miles across the broad American continental shelf, and cascaded over its edge. Those rivers cut the submarine canyons like those of the Hudson and Delaware so prominent on the sea floor today.

The Bahamas were two vast islands—Grand Bahama (with Abaco) in the north, greater Andros extending all the way to Long and Ragged Islands in the south—and several disjunct, smaller islands to the southeast. Cay Sal Bank, between the Florida Keys and Cuba, was an island as large as all Everglades National Park is today. Interestingly, however, the Santaren Channel, between Cay Sal Bank and the Bahamas, and the Nicholas Channel, between Cay Sal and Cuba, remained open. Those big islands of the Bahamas were never connected to Cuba, Florida, or Hispaniola, and none of those were ever connected to each other. And the Gulf Stream poured through the Straits of Florida just as it does today.

The climate was different. Cold, dry air of great density (high pressure) flowed off the glacial ice. When it collided with the warm wet air (low pressure) of the tropics—much closer then than ice and tropics are today—tremendous storms resulted. The West Indies were warm and very wet: much rainier than they are today. This increased moisture enabled the rich fauna and flora of tropical Florida to flourish.

We are not sure when the greatest predator, *Homo sapiens*, first appeared on the scene. In western North America, the wholesale slaughter and—soon—extinctions of animals began as much as forty thousand years ago. In the Antilles to the south, similar evidence of animal death and destruction is documented about 3,500 years before present. Wherever and whenever humans appeared, the large, easygoing animals of the Pleistocene, the Ice Ages, were annihilated. Amerindians exterminated far more species of animals than the coming Europeans and Africans have yet managed to wipe out. When Juan Ponce de Leon arrived in the Florida Keys in 1513 there were few living Amerindians. But whole islands were littered with human bones. Apparently, having eaten up most of the easily killed animals, the Caloosas dined on each other. One bone-strewn isle was named Cayo Hueso: "Bone Cay"; we call it "Key West."

The Wurm Glaciation reached its maximum about seventeen thousand years ago. The ice began to melt. By ten thousand years ago most of the great plain to the south and west was inundated. The grand coral reef that had been a ridge on the Wurm plain was once again lapped by the sea. Today it is the chain of the Upper and Middle Keys. Made of Sangamonian reef called Key Largo limestone, it extends far enough south and west to make the southern rim of Big Pine Key and the little Keys flanking Coupon Bight.

All the rest of the Lower Keys are made of, or founded on, oolite. They are remnants of the once-great plateau of the Wurm, and, before that, the quiet sea bottom of the Sangamon. Waters pouring out of the Gulf of Mexico dissected the oolite plateau into relict platforms. Storm-borne drift sand, coral, and detritus collect in ridges, called *berms*, along the edges of the lozengate platforms of oolite. This loose, sedimentary material traps rain water and can house narrow *lenses* of fresh water in its low dunes and ridges. In general, oolite is much less pervious to water than is Key Largo limestone. Sea water moves laterally through the Key Largo limestone far more readily than through the oolite. Fresh water, from rain, which enters the oolite is readily trapped there, within the *lithic* or rocky *lens*. These big lenses of fresh water provide for species that survive in the Lower Keys, but not in the Middle and Upper Keys: lenses in Key Largo limestone are small. The alligator is a fine example.

The history of the Florida Keys since the end of the Wurm Glaciation is the subject of impassioned debate. There are two fundamentally distinct views. First, the more or less orthodox view is that sea level has risen steadily, albeit at different rates, and stands today higher than it ever has since the Sangamon Interglacial. Second, the heterodox view advanced by the world-famous geologist Rhodes Fairbridge is that sea level has been a veritable picket fence of ups and downs. The "Fairbridge curves" locate sea level as much as four meters (more than 13 feet) above present as recently as 4,000 years ago. Such a rise would wipe out virtually everything living on the land of the Florida Keys, because land would be reduced to a few tiny atolls at Key West and Key Largo. These could support nothing more terrestrial than a few mangroves, some salt grasses, sea birds, and maybe diamondback terrapins and mangrove water snakes (but *not* crocodiles).

If Fairbridge is correct, virtually the entire fauna and flora of the Keys has colonized within the last 3,500 years. Furthermore, *all* colonizations had to be across water, at least to the Lower Keys: sea level never dropped low enough in the meantime for animals to walk all the way.

Fairbridge often used midden sites—the shell heaps (with other discarded junk) of Amerindians—to indicate previous sea levels. This requires assuming that these people ate their seafood at sea level. If I make the Steinbeckian assumption—that they were basically like me—I cannot make the Fairbridgian one. I sometimes eat my seafood at the edge of the sea (beach party), but more often take it home (two to ten meters above sea level, depending on which home). I sometimes dump the shells where I eat, and sometimes haul them or have them hauled to a more remote site, often farther above sea level.

Another explanation has been suggested: land level may have been much higher in the recent past; erosion of the limestones may be rapid. If, for example, the land was four meters higher 4,000 years ago, then the Keys could have retained their fauna and flora, just like today's, right through Fairbridge's rise. The implied rate of erosion would have to be about a meter every thousand years.

To be sure, limestone does erode quickly under the dissolving influences of rain water and the acids released from decomposition of plant material: detritus. However, sea salt buffers and mitigates this process effectively. Even a scant amount of sea salt, as from wind-blown spray, will work. Even one part per thousand in the fresh water, too little to render it impotable to land animals, is an excellent buffer. Therefore, the most striking erosional effects—solution or sink holes—are found at the highest and most inland sites. We shall return to this point because it is profoundly important to many animals and plants.

There are trees in the Keys 500 years old, such as some pines on Big Pine Key. They seem to be growing now where they first grew, not half a meter (about 20 inches) lower.

I do not doubt continental drift, sea level rise and fall, or erosion: all are real phenomena. What I do doubt is the timing, the schedule of these events, that has been proposed by some geologists. I doubt that geology is an exact science, or that the final word has been heard.

References

Enos, P., and R. D. Perkins. 1979. Evolution of Florida Bay from island stratigraphy. *Bulletin of the Geological Society of America* 90: 59–83.

Fairbridge, R. W. 1974. The holocene sea level record in South Florida. *Memoirs of the Miami Geological Society* 2: 223–32.

Gascoyne, M., G. J. Benjamin, and H. P. Schwarcz. 1979. Sea-level lowering during the Illinoian glaciation: evidence from a Bahama "blue hole." *Science* 205(4408): 806–8.

Hoffmeister, J. 1974. *Land from the Sea*. Coral Gables, Fla.: University of Miami Press.

McPhee, J. 1983. *In Suspect Terrain*. New York: Farrar, Straus, Giroux.

Martin, P. S. 1984. Catastrophic extinctions and late Pleistocene blitzkrieg: two radiocarbon tests. In *Extinctions*, ed. M. H. Nitecki. Chicago: University of Chicago Press.

Morris, B., J. Barnes, F. Brown, and J. Markham. 1977. *The Bermuda Marine Environment*. Bermuda Biological Station Special Publication No. 15, St. Georges West, Bermuda.

Shinn, E.A. 1988. The geology of the Florida Keys. *Oceanus,* Woods Hole Oceanographic Institution, 31 (1): 46–53. Excellent. Provides good diagrams.

Webb, S. D., ed. 1974. *Pleistocene Mammals of Florida*. Gainesville, Fla.: University Presses of Florida.

THE FLORA

If you want to be a success as a field zoologist, get yourself about thirty feet of stout chain. Put a botanist on one end of it and do not let go of the other.

The fauna of the Florida Keys is absolutely tied to the flora. Most animals eat and shelter in vegetation. The vegetation indicates critically important features like substrate type and water salinity. Often, vegetation literally makes the habitat. There are even some negative correlations: sea turtles and some sea birds need a relative absence of vegetation for their nest sites. Our native plant associations in the Keys, especially the hardwoods of the hammocks and the understory of the pinewoods, are among the most species rich in all America. These little, low islets are fairly smothered in the diversity of life. That is except where man has destroyed the natural communities.

A grave threat to plant communities is posed by introduced exotics and invasive escapes. Plants brought from other lands (like animals) are *exotic* by definition. Some were brought as food producers (mangoes, bananas, citrus), others as ornamentals (bougainvillea, oleander, flamboyant), and some for useful products like wood and fiber (century plant, Australian pine). The introduction of a few, like Brazilian pepper (*Schinus terebinthifolius*) and punk tree (*Melaleuca quinquenervia*), seems to have been purely malignant.

Little native wildlife lives in the introduced exotic flora, but introduced exotic wildlife often does well there. As with the plants, exotic animals pose a threat to natives. They provide needless competition, vector diseases, and may be major predators. Plants and animals go together, quite literally.

The major plant associations of the Keys are four:

1. *Mangroves.* Three unrelated species are the dominants of this association. The red mangrove, *Rhizophora mangle,* may attain good tree size, but often exists naturally over large expanses as a low shrub—"spider mangrove." This species is the most widespread and is characterized by grand, looping prop roots. The black mangrove, *Avicennia germinans,* grows to be a stately tree. It puts up hundreds of skinny, fingerlike *pneumatophores* from its roots. These make a tall, but sparse, carpet-pile-like appearance on the mud flats. The white mangrove, *Laguncularia racemosa,* likes lower salinities and usually more inland sites. It may have short, dense prop roots, like a corn stalk.

2. *Transition Zone.* This is usually an open savannah habitat studded with buttonwood, *Conocarpus erectus*, its dominant tree. Buttonwood is really as much a "mangrove" as its close relative the white. The transition zone is a periodic wetland, inundated by flood tides. The ground cover of transition zone (virtually absent in mangroves proper) is of great importance to wildlife. Here are meadows of grasses such as Key grass, *Monanthocloe littoralis;* spike grass, *Disthichlis spicata;* and cord grass, *Spartina spartinae.* Here are sedge meadows of *Fimbristylis castanea*, and rush meadows of *Juncus* spp. And there are great swales of saltwort, *Batis maritima.* Where transition zone lies seaward of high land, it can be incredibly rich feeding grounds for deer, rabbits, rice rats, and cotton rats—and hunting grounds for raccoons, bobcats, hawks and owls, and big snakes.

3. *Pinelands.* The higher oolite platforms of the Lower Keys often support a forest-fire-dependent plant association dominated by slash pine,

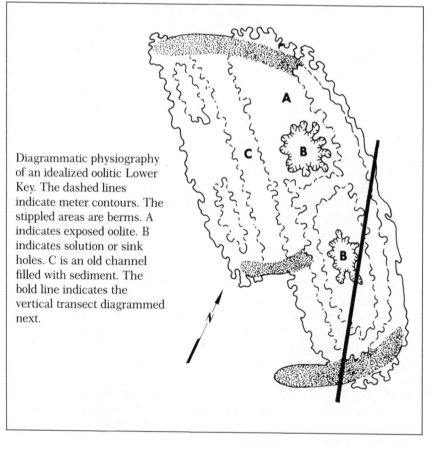

Diagrammatic physiography of an idealized oolitic Lower Key. The dashed lines indicate meter contours. The stippled areas are berms. A indicates exposed oolite. B indicates solution or sink holes. C is an old channel filled with sediment. The bold line indicates the vertical transect diagrammed next.

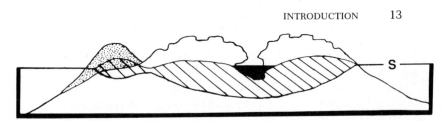

Diagrammatic vertical transect through part of an idealized oolitic Lower Key. The sand berm is stippled. S indicates sea level. The fresh water lens is hatched diagonally. Open fresh water, in the sink hole, is black. Within the sand and rock there is no turbulence to cause mixing, so fresh water floats on and displaces salt water, which is much more dense. Mosquito ditching can destroy a fresh water lens by letting salt water penetrate its surface.

Pinus elliottii densa. These little pines live 400 or 500 years; their wood is so dense it bends nails and so heavy it sinks in water. Beneath them grow many species of palms, including the lovely and elegant silver palm, *Coccothrinax argentatus,* and ferns. There are dense brakes of bracken, *Pteridium aquilinum*. From wet pockets soar magnificent leather ferns, *Acrostichum* spp. The pinelands are primary home for the Key deer, and good habitat for many other species. Pinelands exist on land over well-developed fresh water lenses.

4. *Hardwood Hammock*. This is the climax community of the Keys. That means it is the ultimate stage in plant ecological development: nothing would—in a state of nature—replace it. By accumulation of drift sand, coral, debris, and detritus, mangroves may be succeeded by transition. Moving inland, transition is succeeded by hardwood hammock. In the absence of periodic forest fires, even pineland gives way and hardwood hammock succeeds.

This is the richest zone in tree species in the continental United States. About 120 native species occur, sometimes over 80 in a single hammock. Given that these hammocks are very small, insular areas, on small islands, this species diversity is a notable exception to standard biogeographic rules. Theory has it that species diversity diminishes as one proceeds out peninsulas. Islands, theory claims, should have far lower species diversity than comparable mainland areas. Yet, here we have tiny isles at the remote end of a long peninsula supporting a species diversity of trees far greater than that of the mainland portions of northern Florida.

The reason is overwater colonization from the Bahamas and Antilles. Biogeographic theory would have us believe that colonization goes the other way: from mainlands to islands. More than 60 percent of all the flora of the Keys came across water from the south and east. Considering tree species alone, more than 90 percent of the species are classically West Indian. About 10 percent of the total flora is endemic, most derived from Antillean ancestors.

To the early human colonizers from Europe, this diversity was a source of wonder and a great economic boon. They harvested the mighty mahogany (*Swietenia mahagoni*) and lignum vitae (*Guaiacum sanctum*). They loathed and feared the toxic manchineel (*Hippomane mancinella*) and poisonwood (*Metopium toxiferum*). They used many trees for medicine, like the stoppers (*Eugenia* spp.), and learned to poison fish with Jamaica dogwood (*Piscidia piscipula*).

The hardwood hammocks are a source of wonder today to those who would take the time to wander through them with an expert identifier. Nowhere else in the United States is there anything remotely like these dense ecosystems. They have fed and housed and shaped the unique ways of the wildlife.

References

Long, R. W., and O. Lakela. 1971. *A Flora of Tropical Florida*. Coral Gables, Fla.: University of Miami Press.

Ruttenber, J., and A. H. Weiner. 1978. *Florida Keys Hardwood Hammock Atlas*. Florida Cooperative Extension Service and National Audubon Society, Tavernier, Florida. (This is a rare publication, not in most libraries; it and the following, which is the descriptive text for the atlas, can both be seen and studied at Key West Garden Club, West Martelo Tower, White Street and the Atlantic Ocean, Key West.)

Weiner, A. H., and K. Achor. 1979 et seq. *The Hardwood Hammocks of the Florida Keys*. 2d ed. Key West: Florida Keys Land Trust.

Scurlock, J. P. 1987. *Native Trees & Shrubs of the Florida Keys*. Pittsburgh: Laurel Press, and Key West: Florida Keys Land Trust.

Stevenson, G. B. 1969. *Trees of Everglades National Park and the Florida Keys*. 2d ed. Homestead, Fla.: Everglades National Park.

Tomlinson, P. B. 1980. *The Biology of Trees Native to Tropical Florida*. Allston, Mass.: Harvard University Printing Office.

THE FAUNA

It would be an interesting study . . . to see how far North American species have adapted themselves to the West Indian flora, and how far they have varied under its influence.

L. F. DE POURTALES, 1877

If the flora is largely derived from the Bahamas and the Antilles, to the south and east, then the fauna comes as a surprise. The land animals of the Keys mostly came down from continental North America. It is this very juxtaposition of a North American fauna in an Antillean flora that makes the Keys so interesting. Of course, that juxtaposition is an oversimplification: many animals in the Keys came from the Antilles or Bahamas. The generalization, however, holds.

The fauna provides strong evidence of solid land connections. Groups like ungulates (deer), carnivores (raccoon, bobcat), and lagomorphs (rabbit), are notoriously poor overwater colonizers. Their ancestors probably had to walk (or hop) here. It is the faunas of coastal islands that provide the strongest evidence against the Fairbridge curves of sea level rise and fall since the Pleistocene.

The most often asked question about island animals is, "How did they get there?" For any given species, the method and route may have been unique. In outline form, here are the most obvious possibilities:

  I. *Overwater waif dispersal.* They may have flown (fruit bat) or swum (crocodile) or been carried passively, clinging to floating vegetation (reef gecko). This category divides temporally:
  A. Pre-Wurm, across the Sangamonian sea
  B. Post-Wurm, where water is today
  It also divides spatially:
      1. Bahamian
      2. Antillean (largely Cuban)
      3. North American (largely mainland Florida)
 II. *Stranded populations.* The ancestors of these animals came across dry land during the Wurm Glaciation. Post-Wurm sea level rise has stranded them where they live today. Classic examples are Key deer, alligator, and indigo snake. All of these came from North America.
III. *Composites.* Some animals are adept at making short crossings, but never seem to make long ones. Their original ancestors were probably stranded, but they are able to move about between even fairly remote Keys—over water—today. Examples are raccoons and rattlesnakes.

The presence of any given species on any particular Key will depend on two major conditions: environmental and historical. If colonizers reached the site of the present Key, either over water or over land, were the ecological components present to sustain a population? These components might be fresh water to drink, food plants or prey animals to eat, and shelter sites. Then, just because a species *could* survive on a particular Key does not mean it necessarily ever got the opportunity. History is chancey, and the absence of a species on an island should not be surprising.

Once having established a population in the Keys, the colonizers find themselves facing novel problems not met at home. It may be drier, or hotter, or more open, or the food sources may be different. This last is an obvious problem for a North American animal that has moved into an Antillean plant community. These kinds of new stresses on the population are the selection pressures that bring about evolutionary change and the origin of new species.

Evolution by means of natural selection is brutal. Most colonizing populations quickly become extinct. Few are prolific enough to withstand the grim reaper of selection. The vast majority of all offspring produced must die or fail to reproduce themselves. Only a few well-adapted variants survive and breed. Fecund, prolific species have the best chances of responding in time to novel or changing environments.

For the few survivors, however, the result can be very rapid evolution to new species or subspecies. The Florida Keys have been a most active theater for the evolution of new forms—all in the last few thousand years, since land first broke the surface of the falling Sangamon sea.

Among native mammals, most are endemic at species or subspecies level. All but one, the fruit bat, are of North American origin.

Among native reptiles, about 20 percent are of Antillean or Bahamian origin. The figures are imprecise, because controversy exists over whether some forms are native or were introduced by humans. Evolution here has been slower; about ten percent are endemic subspecies.

Among amphibians, all but three (70 percent) are stranded North Americans. At least two of those three are introduced; one might be native. Described and accepted endemicity is nonexistent even at subspecies level. Some populations seem quite distinctive, however, and biologists have generally grossly neglected the amphibians.

Among birds, about ten percent crossed water from the Bahamas or Antilles. No endemic species or subspecies are recognized.

Invertebrates are simply too little studied to support generalizations. Many, like land crabs and tree snails, are of Antillean origin. Remarkable endemics also occur, especially in the well-documented groups like tree snails and butterflies.

All of this brings up the question of how animals are classified. I refer to animals as species or subspecies as though I clearly knew the difference. I call some animals by generic names that are different from those most of my colleagues use, or that usually appear in popular books. Why? What is it all about? What is a species? a subspecies? a genus?

Lots of people are bored by classification schemes and names. If you are one of these, please just skip this section and go right on to the species accounts. For me, animal classification is the most important aspect of biology because it utilizes all the available information we can glean about the animals involved: anatomy, biochemistry, physiology, behavior, distribution, and ecology. Since I care so deeply, and since the Florida Keys provide such a classic array of examples, I will give some space to these issues and questions.

The basic scheme I use is the one followed by most systematic biologists or taxonomists. It was invented by Linnaeus and published in 1758. He called it the *Systema Naturae,* the System of Nature. In this hierarchical system animals (and plants) are grouped together on the basis of resemblances in anatomy. Each lower category in the hierarchy reflects greater and greater resemblance. This system proceeds all the way from the kingdom to the genus (plural genera): a group of true look-alikes. Then, suddenly, come species. For the first time in the system, animals are split apart within the genus on the basis of differences in detail.

Each kind of animal in the system *must* have two names: first its generic name, capitalized; second its species name, *always* lower case. If the species breaks up into geographical segments that meet several other special criteria (noted below), there may be a third name: the subspecies. If so, some one population or subspecies—the one first named—must be the *nominate* form: its subspecies name is identical to its species name. For example, *Procyon lotor lotor,* the raccoon of the northeastern United States.

Charles Darwin revolutionized the system. The resemblances that had been used to group animals in the hierarchy were often real evolutionary relationships. Sometimes, however, the resemblances were spurious. Animals quite unrelated to each other often evolved similar structures: biologists could be fooled. Bats and birds both have wings, but their respective wings have evolutionary pathways and origins irrelevant to each other. Now systematic biology became the product of evolutionary studies and the lid was off. Anything anyone could find out about the animals might shed new light on their true relationships, and thereby improve the system.

It is important to clearly distinguish the container from the thing contained (remember James Thurber's eloquent exposition relative to "I hit him over the head with the whiskey"?). The containers are called categories. The things contained are called taxa (singular: taxon). Each kind of organism belongs to its own set of taxa in the system of categories. For example:

CATEGORY	TAXON
Kingdom	Animalia
Phylum	Chordata
Class	Mammalia
Order	Carnivora
Family	Procyonidae
Genus	*Procyon*
Species	*lotor*
Subspecies	*auspicatus*

That is the Key raccoon. The Northeastern raccoon is just the same all the way to subspecies. A bobcat diverges at the category Family. It belongs to the Family Felidae, with the rest of the cats. Animal family names always end in -idae, in their latinized form. All proper, latinized taxa are capitalized, *except* species and subspecies (the ones diagnosed by differences instead of defined by resemblances). Generic, specific, and subspecific taxon names are *italicized* or underlined, like the foreign words they usually are.

Other systems of classification are frequently promulgated. Two have garnered numerous vocal adherents during my years as a systematist. One, called "numerical taxonomy" or "phenetics" called for utilizing only quantitative characters in classifying animals, and, furthermore, in discarding all subjective weighting of those characters. For example, to believe that the number of bones in the distal portion of a deer's foot is more important than is the number of tines on its antlers is subjective and disallowed. The end products of phenetic systems were not the species and subspecies in which I am interested. They were *"phenons"*—collections of quantitatively similar individuals. Often, males and females of the same biological species ended up in different phenons; for example, with deer: antlers two versus antlers zero. A cat with the genetic anomaly for multiple toes would not be classified with its litter mate who had the normal complement.

Another system has been called "Hennigian systematics" or "cladistics." It was invented by the German anatomist Willi Hennig in about 1950. Here the fundamental dogma is that only the actual time of evolutionary separation of lineages should be used to rank those lineages, taxa, in categories. This system would not admit widely differing evolutionary rates in genetic material (DNA), or anatomy, and ignored resemblances inherited from common ancestors. Only the actual temporal propinquity of common ancestry mattered. By this system what I call the Class Mammalia became a mere genus of fishes. All island populations isolated at the same time by Post-Wurm sea level rise would have to be given the identical categorical rank. If any can be called a new species, all must be.

I still use the *Systema Naturae* as modified by evolutionary biologists following Darwin. The question, then, is how does one determine which taxa belong in which categories?

I group animals together according to their real, evolutionary resemblances right down to genus, just like most of my colleagues. I insist on some definite principles: each taxon, at each categorical level, must be defined in a way that absolutely and unequivocally identifies its members. And some definitive distinctions must be hard anatomy. I will use all the other kinds of data I can get to clarify relationships and indicate real, evolutionary resemblances. But in the end the definition of the taxon must be one that enables me to recognize its members in the fossil record. I want paleontology, the study of ancient life, to remain relevant and comparable to neontology, the study of living things.

The genus, an animal species' first name, is the last of the imaginary taxa. It is a word that must be unequivocally defined: "usually" can never appear in the definition of a valid genus. Furthermore, intergeneric hybrids cannot exist. How could they? As soon as two sorts of animals, nominally placed in separate genera, mate and produce viable offspring, the integrity of the generic definition has been destroyed: an animal with intermediate, equivocal characteristics exists. Or, viewed another way: any two species that are so closely related that they can interbreed and produce viable offspring are as closely related as species can possibly be and still—arguably—even be different species. Horse and donkey, bobcat and house cat, dog and fox, ratsnake and pine snake are all examples: they cross breed, but only rarely or in artificial circumstances like captivity.

The degree of anatomical difference between interbreeding species, like the degree of anatomical difference within a variable species, can be used as a fine indicator for generic limits in other groups where the reproductive data are missing. If you consider the spectrum of morphological variation within a species like the dog (*Canis familiaris:* from peekapoos to great danes), putting foxes (*Canis vulpes*) in the same genus seems quite unremarkable.

I do not accept as valid those genera that have been defined on the basis of secondary sexual characteristics such as plumes, color patterns of males, antlers, accessory structures of the copulatory organ(s), and so on. These basically qualify as ornaments, although some may be of great importance as behavioral and/or mechanical isolating mechanisms for closely related species. They are unacceptable generic characters for two reasons. First, they are present only in one sex, therefore falling far short of the consistency requirement. Second, they are just the sorts of features natural selection tends to make most different in the most closely related species. Male bird plumages provide the most familiar examples. Species with males that look completely different may have very similar females. Character divergence has been selected for because hybrid mistakes are evolutionary dead ends. Striking differences in secondary sex characters may indicate extremely *close* relationship.

I want all valid genera to be readily recognizable. I cannot ask that for species: many are so similar only an expert recognizes them. But genera, no. One should be able to identify every specimen that comes to hand to its proper genus, given only a list of the definitions.

Species are lineages of individual organisms that evolve together as a unit. They are real. At any given time one can sally forth and determine the limits of a given species (it may not be easy, but it can be done). The best way for a group of real animals to maintain their evolutionary unity is by freely exchanging genetic material: interbreeding. However, one must not put too much weight on the mere fact of interbreeding. Remember the horse and the donkey: they *can* interbreed to produce a mule. Despite this they go right along their own different and separate evolutionary pathways. The existence of mules in no way unites horse evolution with donkey evolution.

Subspecies are geographic variants within a more wide-ranging species. They are unified with the other subspecies that make up the whole species by merging at their edges. The rule for recognition of subspecies is that three out of four individuals must be correctly identifiable *without* recourse to locality data. It is not fair to diagnose a subspecies on the Torch Keys as being "just like its relatives on Big Pine, except that it comes from the Torch Keys." There has to be a recipe—a diagnosis—that identifies the vast majority of specimens by their physical characteristics.

Island forms present great problems. If they are not absolutely distinct, but more than 75 percent can be identified, I call them subspecies. What if they are 100 percent, absolutely distinct—like the Key deer and a mainland white-tail? What are they then?

Most people say, "Why not just put some together and see if they interbreed?" Well, that is not good evidence. Under artificial circumstances all sorts of species can be cross-bred. Herpetologists and mammalogists put some interbreeding animals in separate genera (e.g., ratsnake and pine snake, fox and dog, or bobcat and house cat), while ornithologists have interbreeding *families*—like "Phasianidae" (pheasants and chickens) and "Meleagridae" (turkeys).

In deciding the status of each and every island form, all of the evidence must be sifted and weighed. Even then, I may evaluate the number of a snake's ventral scales as more important in classification than the number of millimeters it is long. A colleague may simply disagree. Island forms are often on their way to becoming species, but may not have quite achieved full independence, their own evolutionary unity, at this time. In arguable cases, I state why I have classified the form as I have.

The animals of the Florida Keys remain little known. In discussing their relationships and geographic ranges I am often making surmises based on scant evidence—sometimes a single specimen. Therefore it is often necessary to cite individual specimens housed in museums. This is done by museum acronym and individual specimen number. Following are the museum acronyms: MCZ, Museum of Comparative Zoology (Harvard); YPM, Peabody Museum of Natural History (Yale); NMNH, National Museum of Natural History (Smithsonian); AMNH, American Museum of Natural History (New York); FSM, Florida State Museum (Gainesville).

Those who wish to explore the Keys should remember that *all* of the fauna of the Keys is legally protected. Much of the Keys is in National Wildlife Refuge or National Park lands where *all* wildlife is protected. Federal law also explicitly forbids the taking or molesting of many species listed as rare or threatened. Key deer, alligator, crocodile, Key woodrat, Schaus swallowtail, silver rice rat, and all sea turtles are just some examples (the lists change). State law further protects many species as state-threatened or endangered. Red rat snakes, indigo snakes, mangrove terrapins, and tree snails are a few examples. A state permit is required for any sort of specimen collecting; it is difficult to get unless you are a scientist with credentials and a specific research objective.

Absolutely NO hunting is allowed in the Florida Keys. Down here it is not hunting—it is poaching. Jack Watson is gone, but his spirit is still with us.

"Just how do you figure to arrest us?" one poacher said, shotgun draped over his arm. "There's five of us and only one of you."

Watson's hand came out from behind his back with the .357 magnum. "You're wrong," Jack said. "There's six of me."

References

Buckley, P. A. 1982. Avian genetics. In M. Petrak, ed. *Diseases of Cage and Aviary Birds*, 2d ed., pp. 22–110. Philadelphia: Lea & Febiger.

De Pourtales, L. F. 1877. Hints on the origin of the flora and fauna of the Florida Keys. *American Naturalist* 11: 137–44.

Gray, A. P. 1972. *Mammalian Hybrids*. Farnham Royal, England: Commonwealth Agricultural Bureaux.

Simpson, G. G. 1961. *Principles of Animal Taxonomy*. New York: Columbia University Press.

Van Gelder, R. G. 1977. Mammalian hybrids and generic limits. *American Museum Novitates* 2635: 1–25.

MAMMALS

. . . their enormous range in vagility makes mammals particularly apt subjects for biogeographic research.

L. R. HEANEY AND B. D. PATTERSON, 1986

Mammalia is the only Class of chordate animals in which the lower jaw consists of a single bone—the dentary—on each side. The small bones that make up the proximal end of the lower jaw in other bony chordates have either been lost (angular, coronoid, and splenial) or have moved into the middle ear to form the auditory ossicles. The mammalian middle ear invariably contains three small bones, the maleus, incus, and stapes. Only the stapes (or columella) is shared as an ear bone with non-mammal groups.

This feature enables recognition of mammals in the fossil record. Two other characters of soft anatomy support the unity of the group. Of the great arteries coming out of the heart, called "systemic trunks" or aortas, only the left remains. A muscle wall, the diaphragm, separates the body cavity into an anterior, pleural, portion and a posterior, peritoneal, portion. No other animals have a diaphragm. The presence of hair or fur and the state of being "warm-blooded" have no value as definitive characteristics of Mammalia, despite popular opinion. Many mammals, such as dolphins, lack hair; hair is not readily separable from nonmammalian structures, such as the vibrissae (whiskers) of birds. Many mammals, especially small tropical forms, are quite as incapable of maintaining a high, stable body temperature as any reptile. A number of living reptiles (pythons, leatherback) are "warm-blooded," as are bluefin tuna fish.

The famed mammary glands are an interesting feature. They are secondary sexual characters developed in females: ornaments indeed. In all mammals except *Homo sapiens* they are quite inconspicuous except when engorged with milk and functioning to feed the young. Monotremes (the egg-laying mammals) do not even have nipples. The spectacular enlargement of the region around the mammary glands, and of the nipples, in nonlactating humans is a unique feature.

Since we ourselves are mammals, one might expect us to take a great interest in this Class but, generally, we seem not to. Myriad amateur societies focus on birds, reptiles and amphibians, mollusks, butterflies, fishes, and even fungi, but there is hardly a "mammal club" on Earth. From grand beginnings in the nineteenth and early twentieth centuries, mammalogy has declined considerably.

I believe the problem lies in recruitment. Most other popular groups of animals (and plants) have had their champions who took their cases to the people. John James Audubon and Roger Tory Peterson (birds) and Roger Conant, Hobarth Smith, and Archie Carr (amphibians and reptiles) are examples. High-quality field guides and popular books became available to the public; these works pointed out what was *not* known, and how to discover it. Amateurs and professionals alike could find out when they had encountered something new, and could set out to explore and search for the unknown. Mammalogy—right here in the United States, and over the world generally—is a wide-open field for alpha-level work and discovery, but precious few books offer any assistance.

There are no good mammal guides, except a couple that cover selected groups (like whales). To find out anything about most mammals, one has to go directly to the technical literature. Even then, the taxonomy of mammals is a mess. Numerous nominal subspecies are retained for lack of the sort of investigations that would sink them. Other forms remain undescribed. I know of at least two undescribed mice on East Coast islands that would probably qualify as subspecies—one is likely a full species. Unfortunately, neither of these mice has been thoroughly studied. Nobody cares about a new species of mouse, it seems. If it were a new species of bird, it would make the front page of the *New York Times*.

In the following accounts I have omitted several species now extirpated from the Keys (black bear), known from single specimens undoubtedly escaped captives (a hare, genus *Lepus,* YPM 8926), and at least two introduced species that are prospering. These are the red-bellied squirrel, *Sciurus aureogaster,* brought from Mexico to Elliott Key, northeast of Key Largo, and the rhesus macaque, *Cercopithecus mulattus*. This monkey, imported from India, is raised on Loggerhead Key ("Key Lois"), south of Summerland, and on Raccoon Key, north of Sugarloaf. These stocks are owned by Charles River Laboratories and bred for medical research. Occasionally, a monkey gets off its devil's island and reaches some other key. One is said to reside on Lower Sugarloaf. A hurricane might spread these animals all over the Keys. They are extremely destructive to vegetation and potentially dangerous to people.

An account of most of the species I have omitted is given in the reference by Dr. Layne, cited below.

References

Hall, E. R. 1981. *The Mammals of North America*. 2d ed. 2 vols. New York: John Wiley & Sons.

Heaney, L. R., and B. D. Patterson. 1986. *Island Biogeography of Mammals*. New York: Academic Press.

Lawlor, T. E. 1979. *Handbook to the Orders and Families of Living Mammals*. Eureka, Calif.: Mad River Press.

Layne, J. N. 1974. The land mammals of south Florida. *Memoirs of the Miami Geological Society* 2: 386–413.

Schwartz, A. 1952. *The Land Mammals of Southern Florida and the Upper Keys*. Ann Arbor, Mich.: University Microfilms.

Watson, L. 1981. *Sea Guide to Whales of the World*. New York: E. P. Dutton.

KEY DEER
(DAMA CLAVIA)

With the development of human populations in the mid- to late-1800s . . . clearing and harvest of the wooded areas . . . resulted in Key deer being viewed as a pest, and also encouraged use of it for food, . . . reaching annihilation levels by the early 1930s. . . . Hence, by the 1950s it was suggested that the population was as low as 25 deer. . . . And in 1951, Jack C. Watson, who devoted nearly 25 years of his life to the deer, was employed and endowed with both state and federal enforcement authority.

J. W. HARDIN, W. D. KLIMSTRA, AND N. J. SILVY, 1984

When I first took up part-time residence on a back-country key in 1973 I met the Poacher. He was an amiable fellow, getting on in years, and pretty much retired. He still popped white-crowns out of the pigeon plums if he could retrieve them without walking more than thirty paces from his porch, and he still thought it a fine thing that other folks would set and bait lobster traps for him to visit on moonless nights.

He still had an ample supply of his Torch hounds, too. Torch hounds looked sort of like beagles or small redbones but had the freakishly short legs of a basset or dachshund. The Poacher had developed his own breed for rooting game out of the dense low hammocks and mangroves. No Key rabbits were left on any of the four big keys in the complex where the Poacher lived, and, yes, he reckoned between him and his hounds they had all been eaten up some years before. But there were quite a lot of Key deer.

"How come," I asked, "if you've been poaching here so long, with these fine, special dogs of yours, there are still so many deer around?"

He groaned aloud. "It's that Jack Watson feller," he said.

One night up on Big Torch the Poacher was working with his brand-new sealed beam spotlight. Watson stepped out from behind a stopper sapling and had the Poacher dead to rights.

"I didn't have no dead deer, or anything," he said. "Just the light. 'Course he knew exactly what I was doing wasn't shining rattler's eyes."

On that night Watson struck a very simple deal with the Poacher. Watson would patrol the road every once in a while in the evening. Not only would he never see a spotlighter working, he would have to see deer on a regular basis—reasonable numbers of deer, too. If he—Watson— didn't, the Poacher said, Watson was simply going to come and shoot him—the Poacher—and every one of his short-legged dogs.

"It's been a burden and a hard job of work ever since," the Poacher said. "See, in dry spells those deer will wander off these keys and go over to Big Pine to get water. I can't let that happen. I got to get out and muck out that water hole over yonder all through the dry weather so there'll still be deer

Key deer, *Dama clavia*. Big Pine Key.

Joseph J. Oliver

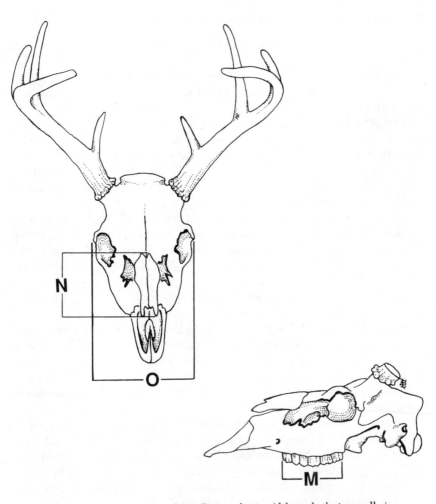

Diagram of the skull of the Key deer, *Dama clavia*. Although their small size impresses most observers, their differences from ordinary white-tails, *Dama virginiana*, are far more profound. Their skulls are as broad as northern white-tails. The measurement 0, for orbitofrontal width (frontojugal of Hall, 1981), is 90–119 mm, not significantly different from that of a hundred white-tails from mainland Florida to Maine: 90–118 mm. The teeth of Key deer are strongly distinct in many ways from those of white-tails. Just the simple measurement M, the total length of the molariform row, provides no overlap. The largest Key deer molariform row is 66.4 mm, in the type-specimen. The smallest white-tail I can find is 72 mm, although Hall (1981, p. 1007) gives 68 mm. The nasal bones of Key deer are very broad but quite short. I find their maximum length, N, to be highly diagnostic in ratios with other measurements.

here for that Watson feller to see. He surely knows how to get his work done."

In 1980 Dr. Willard D. Klimstra and his colleagues completed a recovery plan for the Key deer. It is a fine piece of work and its provisions should be implemented.

The actual range of the species is larger than they indicate. Key deer extend west beyond Lower Sugarloaf to some of the southern Saddle-bunch. In the east they have it right: to the Johnston Keys north of Little Pine. There are today no deer on Johnston Key north of Sugarloaf, but this key is entirely in wildlife refuge holding, has fine habitats, and permanent fresh water. It could support a few deer.

Throughout their work, Klimstra et al. provide a clearly Lamarckian view of Key deer evolution and genetics. This harks back to the fifties, when detractors claimed Key deer were just "stunted" white-tails, produced by poor food, little water, and hard times: not really a different kind of deer at all.

Not for a moment do I believe this. The differences between a Key deer and a mainland white-tail are far more profound than just size. The Key deer has a broad frontal region, short, broad nasal bones, and a short molar tooth row. Northern white-tails, liberated and surviving on Lignumvitae Key, have been there for generations, but are still very much northern white-tails. Key deer also have disproportionately long tails and, though variable in color, do not have a summer red and winter grey moult like mainland deer.

Of course all of these differences *could* be environmentally induced. To determine this one need only raise Key deer in captivity, feed and water them well, and see what they grow up to be. Predictably they would grow bigger than wild deer, but the proportional distinctions should hold. I accord Key deer rank as a full species because they fit all the criteria given by Hall (1981, cited for mammals in general, above).

I do not approve of feeding wild animals, and support the Fish and Wildlife Service for prohibiting this practice by tourists (now also a misdemeanor under state law). However, handouts of food will not alter Key deer genes or taxonomic status. Numi Goodyear, of rice rat renown, suggests the perfect Lamarckian solution: make those who would feed the deer squat down and hold the food close to the ground. Then the deer would have to hunker down to eat it. They might get fat, of course, but they would stay short.

Lamarck was the biologist who, before Darwin, believed in evolution by the inheritance of acquired characteristics. That, somehow, giraffes got long necks by stretching to reach high, leafy branches. But this theory simply does not work. Actually, what Klimstra et al. are more nearly suggesting is "Geoffroyism"—an adaptive response to an external stimulus. All of this is clearly compared and contrasted to Darwin's view by Professor Mayr in the first few pages of his book cited below.

Not only do I grant Key deer full species status, I return them to their correct generic name, *Dama*. They have commonly been called "*Odocoileus virginianus clavium*" (as in Barbour and Allen, 1922), and most biologists are unaware that *Dama* is the correct name. Baker (1984) provides a succinct description of the nomenclatorial issues, but he neglects two critically important points: First, the fallow deer, *Cervus dama*, cannot be placed in its own genus (whether "*Dama*" or "*Platyceros*") because the putative generic distinction was its palmate antlers. These are present seasonally in males of just *one of the two subspecies* of fallow deer. The Europeans who petitioned for validation of "*Dama*" for their deer had no rational basis for doing so. Second, and most compelling, I here paraphrase the great mammalogist, Dr. E. Raymond Hall: stability of nomenclature will be most rapidly and lastingly achieved by *obeying* the rules, not by suspending or breaking them. *Dama* is the *right* name for our deer. It is as much a moral issue as zoology provides.

In my view, the fate of these wonderful little deer depends entirely on land acquisition. The deer need room to range and feed. They need fresh water to drink, and that is permanently available only on the higher keys in pinelands and hardwood hammocks. Dogs and automobiles take a terrible toll on them. Single-family residences, sprawled one to the acre over Big Pine, No Name, and the Torch Keys will doom most of them outright. The government should have bought the land thirty years ago when it was plentiful and cheap. It is not too late, but our investment will be vastly greater. Of course, we now know of several other unique forms that would benefit—perhaps be saved—by such land acquisition too: silver rice rat, Key rabbit, Key mud turtle, and Key ringneck snake, to name a few.

The Key deer must have first colonized this area during the Wurm glacial maximum and been stranded in the Keys by Post-Wurm sea level rise. Selection pressures were severe, and evolution rapid. As a new species all its own, the Key deer probably has clear evolutionary independence from mainland white-tails going back less than ten thousand years.

References

Baker, R. H. 1984. Origin, classification, and distribution. In L. K. Halls, ed. *White-Tailed Deer: Ecology and Management*. Washington, D.C.: Wildlife Management Institute, pp. 1–18.

Barbour, T., and G. M. Allen. 1922. The white-tailed deer of eastern United States. *Journal of Mammalogy* 3: 65–78.

Carey, J. 1987. Trouble in paradise. *National Wildlife* 25(6): 42–45.

Hardin, J. W., W. D. Klimstra, and N. J. Silvy. 1984. Florida Keys. In L. K. Halls, ed., *White-Tailed Deer: Ecology and Management.* Washington, D.C.: Wildlife Management Institute, pp. 381–90.

Humphrey, S. R., and B. Bell. 1986. The Key deer population is declining. *Wildlife Society Bulletin* 14(3): 261–65.

Klimstra, W. D., J. W. Hardin, M. P. Carpenter, and S. Jenkusky. 1980. *Florida Key Deer Recovery Plan.* University of Southern Illinois, Carbondale: Cooperative Wildlife Research Laboratory.

Maffei, M., W. Klimstra, and T. Wilmers. 1988. Cranial and mandibular characteristics of the Key deer (*Odocoileus virginianus clavium*). *Journal of Mammalogy* 69(2):403–7. Confirms the distinctness of the species in a complicated way without making the simple taxonomic judgment.

Mayr, E. 1970. *Populations, Species, and Evolution.* Cambridge, Mass.: Belknap Press, Harvard University.

PANTHER
(FELIS CONCOLOR CORYI)

Although some authors have claimed that the panther once inhabited the Florida Keys, earliest writers did not mention it as a part of the fauna. Somewhat special is the Ned Buntline kill of one during the third Seminole War on Key Largo. . . . This record may be as fabricated as the Buntline dime-novels of the western frontier. . . .

JIM BOB TINSLEY, 1970

I could not have improved upon or added to that succinct account until Sergeant Larry Lawrence of the Florida Game and Fresh Water Fish Commission (FGFWFC) told me a panther had reportedly been photographed on Key Largo in 1984. I contacted Robert C. Belden, biological scientist supervisor, FGFWFC Gainesville office, the acknowledged authority on Florida panthers. Belden sent me copies of the correspondence and photo. The animal was unquestionably a house cat, *Felis catus.*

Tinsley provides a good, detailed account of the Matecumbe "black panther" of 1927. The specimen is in the Charleston Museum, South Carolina. It too is a house cat.

These mistaken identities highlight problems that plague those of us interested in the conservation of the magnificent native cat, now federally and state listed as an endangered species. Downing (1984) provides fine documentation of the difficulties and dilemmas. Real panthers, or cougars, are obviously *not* unmistakable animals. So far as we—the scientists—know, they are simply *never* black (see Ulmer, 1941). They are tawny, like the cougar in the Mercury automobile ads. For an excellent color photo see Flowers (1985).

Under contract to the U.S. Fish and Wildlife Service back in 1980, I made a study of the characteristics of the two eastern subspecies of *Felis concolor*. Both differ strikingly from western mountain lions or "pumas" in having broad, inflated nasal bones. These give the animals a "high-brow" appearance: the head is bowed convexly in profile. This is shown well in two of Tinsley's photos on pages 28 and 29. In the Florida subspecies the pelage is short, less than 20 mm at middorsum. The upper canine is massive.

The eastern cougar, *Felis concolor couguar*, never occurred in Florida. It occupied the Appalachian uplands, probably as far south, at least, as the Great Smokies, and as far north as New England and maritime Canada. It occurred as far west as Wisconsin, where it was named *Felis c. "schorgeri."* This subspecies differs from the Florida panther in having long pelage—greater than 20 mm at middorsum—and in having a light, short upper canine. Compelling evidence indicates that this subspecies, officially regarded as extinct, does survive in Maine and New Brunswick.

The Florida panther ranges along the Gulf Coastal plain to eastern Texas (where it is called *F. c. "stanleyana"*) and north in the Mississippi lowlands at least to Arkansas. George Lowery (1974) provided a fine account of Louisiana individuals.

South Florida specimens often show two peculiarities: they may be heavily dappled with white on the nape and shoulders; and they may be "ridge-backed." This latter peculiarity I have never seen in a specimen taken north of, say, Gainesville. The "ridge-back" condition resembles that of the Rhodesian ridge-back hounds: a short patch of middorsal fur is whorled forward—like an elongate cowlick—right along the spine.

I was once asked to testify against R. C. Belden in a case involving the illegal killing of an Everglades panther. The lawyers offered considerable inducements and pointed out that diagnostic skull measurements could be tricky, difficult to make, and arguable.

"Let me ask you three questions about the carcass," I said. "First, you are sure it is a big, brown cat—right? I mean, it's not a dog?"

"Yes," they were certain: it was "a big, brown cat, not a dog." You get into the panther business and you will be absolutely astounded by the number of people who cannot tell a cat from a dog.

"Okay," I said, "does it have little white speckles on its neck?"

"Er, yessir, it does."

"And finally," I asked, "does it have a cowlick on its back: where the fur turns and runs forward?"

"Er, um, yessir—that too."

"Well," I said, "you would be asking me to make a complete fool of myself. There is no animal known to me on Earth that looks like that and is not a Florida panther."

Jack Watson pursued dozens of Lower Keys panther stories over the years. At first he had high hopes, but as time went by he began to joke about the situation. He trailed "panthers" through hammock and swamp. "Claw panthers," he would say. "See, they show distinct claw marks in their tracks."

"I know this panther," he'd say, pointing out a track. "It's Rover. Last week I found Spot, Fido, and Ol' Blue. . . ."

Maybe Ned Buntline wasn't just yarnin'. Maybe he really did shoot a Key Largo panther. There is no reason one could not visit the Key: panthers easily range that far from known Everglades localities. Today, however, with no deer left to eat, a panther on Key Largo would lead a tough life. Panthers do eat raccoons and even smaller game, but deer are their staple.

I cannot improve on R. C. Belden for the last word (*in litt.*, 3 September 1985): "To my knowledge, there has never been any documented evidence of Florida panthers occurring in the Keys."

References

Bangs, O. 1899. The Florida puma. *Proceedings of the Biological Society of Washington* 13: 15–17.

Buntline, N. 1880. My first cougar. *Forest and Stream* 13(24): 994.

Downing, R. L. 1984. The search for cougars in the eastern United States. *Cryptozoology* 3: 31–49.

Flowers, C. 1985. Starting over in the Everglades. *National Wildlife* 23(5): 52–63.

Lazell, J. D. 1981. *Diagnosis and Identification of the Races of Felis concolor in Eastern North America.* Jamestown, R.I.: The Conservation Agency. (Available from the agency, Jamestown, R.I. 02835, for US $10.00; 23 pp., six figures.)

Lowery, G. H. 1974. *The Mammals of Louisiana and Its Adjacent Waters.* Baton Rouge: Louisiana State University Press.

Simberloff, D., and J. Cox. 1987. Consequences and costs of conservation corridors. *Conservation Biology* 1 (1): 63–71. This paper uses the Florida panther as an exemplar.

Tinsley, J. B. 1970. *The Florida Panther.* St. Petersburg, Fla.: Great Outdoors Publishing Co.

Ulmer, F. A. 1941. Melanism in the Felidae, with special reference to the genus *Lynx. Journal of Mammalogy* 22: 285–88.

BOBCAT
(FELIS RUFA FLORIDANA)

Its method of attack is to sneak up on its prey. Sometimes it lies in wait; often it follows trails and runways until it encounters its victim. It eats voraciously and furiously, growling and hissing at anything that interferes with it.

HARTLEY JACKSON, 1961

Sounds pretty dangerous, doesn't it? In the very next paragraph, however, Jackson points out: "Man is the worst enemy of the bobcat, . . . once an individual is located it is readily killed." Nevertheless, the bobcat has done better surviving at the edges of civilization than any other wild cat, except possibly the Asiatic *Felis bengalensis* or leopard cat. The bobcat is almost the coyote of the cat family.

Bobcats still survive on Key Largo. Numi Goodyear found scats in 1985 while doing a study of woodrats and cotton mice. If the wonderful hammocks of north Key Largo are preserved, as I believe they must be, the bobcat will probably hang on as a Keys resident. At one time, bobcats occurred as far down as Lower Matecumbe, as witnessed by USNM 255047. This fine specimen was taken by E. W. Nelson on 20 March 1930. Nelson had just retired as chief of the Biological Survey, which was the forerunner of the U.S. Fish and Wildlife Service. He was down in the Keys collecting raccoons; he will figure prominently in my account of them.

Nelson thought his capture of this big female bobcat on Lower Matecumbe was a lucky fluke. He took her in an unbaited 'coon trap set in a 'coon trail. She was probably following that trail just as Jackson suggests: looking for a plump young raccoon to dine upon. I believe bobcats are now extirpated from Lower Matecumbe and all the other Keys except Key Largo, but I hope someone can refute me. Nelson, writing back to his Smithsonian colleagues, believed that Lower Matecumbe "is as far south on the Keys as the bob cat gets."

What of the constant tales of "wildcats" (not to mention panthers) in the Lower Keys? The question has major biological import for two reasons. First, bobcat populations were surely stranded on the Lower Keys by post-Wurm sea level rise, and several of those Keys are big enough to support populations still (bigger than Lower Matecumbe). Second, what was the natural, native predator of the Key deer?

It is difficult for an ecologist to perceive or admit of a population of large ungulates (and Key deer are large enough) that existed in a state of nature without a natural predator. Could alligators and crocodiles and sharks have been sufficient? Possibly, but I doubt it. The most effective deer predators the world over are cats (except where humans have usurped their role). Without a persistent predator, ungulate populations tend to explode, consume their available resources, and commit ecological, populational, and evolutionary suicide. Like humans, and seemingly unlike elephants and wolves, most ungulates simply have not wits enough to control their own populations. I do not believe *Dama clavia*, the Key deer, would exist today had it not had its own, specialized predator. A form of bobcat would have been perfect.

And that brings up the taxonomic question: what form of bobcat? If the deer has undergone dramatic speciation in the Lower Keys, why would not its predator have done likewise? The predator population would always have been small compared to the prey, so rapid evolution would have been highly probable. Was there an endemic Lower Keys bobcat? Is it extinct?

I believe the answer to the first question is maybe and the second is yes. I am convinced no native wild cat exists in Florida's Lower Keys today. Bones might well turn up in sink holes and midden sites, if archeologists and paleontologists ever got to looking for them. But a living, native bobcat—even in the wilds of No Name or Little Pine—I truly doubt it. If there is a hope of a chance, Little Pine would be the place to search. While you are at it, please relocate Jack Watson's crocodile nesting sites. I can tell you exactly where to look.

I have called the remaining Keys bobcats *Felis rufa floridana*. You will be hard pressed to find that name in any of the literature on bobcats, and harder pressed to justify one component of it if you investigate. The controversial component is *not* the genus *Felis*. Most authorities put this species in the genus "*Lynx*," but that is simply silly. Bobcats and house cats readily hybridize; I grew up with "pet" hybrids owned by friends in Mississippi. The "genus *Lynx*" is justified because it has a short tail and lacks the small upper premolar tooth. Lots of cats—including house cats—have short tails. The Afro-Eurasian species like caracals and servals span the range of tail length possibilities in the genus *Felis* quite naturally (for those who reject Manx cats as "freaks"). The little upper premolar is frequently missing in house cats quite naturally and standardly absent in other wild cat species, such as *Felis yagouaroundi* of our Southwest. So, the putative generic characters do not hold up, and the species readily interbreed.

The controversial issue is the subspecies *floridana*. First described by Samuel Constantine Rafinesque in 1817 (in *American Monthly Magazine*, vol. 21, no. 1, p. 46) from just anywhere in "Florida," the form is instantly suspect. Most of Rafinesque's sojourns into taxonomy were flights of fancy. For a delightful account of this unique "aspiring institution on the . . . American frontier," see Hanley (1977).

As has been standard in American mammalogy, the old names given as full species, like Rafinesque's *floridana,* were reduced to subspecies and quite uncritically retained as such. Modern authorities give the range of *F. r. floridana* as from eastern Louisiana and southeastern Missouri around the Gulf and Atlantic coastal plain to southeastern Virginia, and all of Florida. If they say anything at all about the differences between *floridana* and an ordinary bobcat, *Felis r. rufa,* it is just that *floridana* is smaller and darker. But that certainly is not true, as anyone who simply examines specimens can plainly see.

Smaller might make sense. Bergmann's Rule says cooler-climate populations within a species *should be* larger: a greater volume to surface ratio mitigates heat loss. However, consider Nelson's Lower Matecumbe specimen (USNM 255047): she is as big as any normal bobcat from anywhere in the species range. She measured, in the flesh, 820 mm total length with a 160 mm tail. Her hindfoot was 165 mm. She weighed 9.5 kilos (21 lbs). Males get bigger, but for a female that was a big bobcat.

Darker makes no adaptive sense. Warm-climate individuals—especially on the arid, sun-baked Keys—should be selected for paler, reflective colors; cotton rats, raccoons, and ratsnakes certainly are.

Allen's Rule says that one should expect longer extremities in hotter portions of the species range. This makes sense for cooling and is exemplified by Key cotton rats, raccoons, and probably other species. No one has analyzed bobcat data over the range of species, or—especially—along the axis of the Florida peninsula down into the Keys.

I have listed a plethora of bobcat references below because I believe this species merits an exhaustive study by some aspiring biologist. These are truly beautiful creatures deserving of our attention and conservation.

References

Godin, A. J. 1977. *Wild Mammals of New England.* Baltimore, Md.: Johns Hopkins University Press.

Hamilton, W. J. 1943. *Mammals of Eastern United States.* Ithaca, N.Y.: Cornell University Press. (Greatly superior to the second edition of this book, published in 1979.)

Hanley, W. 1977. *Natural History in America.* New York: Quadrangle, New York Times Book Co.

Jackson, H.H.T. 1961. *Mammals of Wisconsin.* Madison: University of Wisconsin Press.

Lowery, G. H. 1974. *The Mammals of Louisiana and Its Adjacent Waters.* Baton Rouge: University of Louisiana Press.

Schmidly, D. J., and J. A. Read. 1986. Cranial variation in the bobcat (*Felis rufus*) from Texas and surrounding states. *Occasional Papers*, The Museum, Texas Tech: 101: 1–39.

Schwartz, C., and E. Schwartz. 1981. *The Wild Mammals of Missouri*. Rev. ed. Columbia: University of Missouri Press.

Young, S. P. 1958. *The Bobcat of North America*. Harrisburg, Penna.: Stackpole Co.

RACCOONS
(Procyon lotor subspecies)

Study of the Recent material convinces me that P. lotor is the most variable carnivore, and one of the most variable mammals in North America.

Robert A. Martin, 1974

I grew up hunting 'coons. I went on my first 'coon hunt, and shot my first raccoon, in November 1945. That was with Dr. John Minehart of Philadelphia, back when there was still wild land in Montgomery County, Pennsylvania. Raccoons are valuable for their pelts (the one I shot in November, 1945, is at the Peabody Museum at Yale); the meat is excellent—rich and red; and the sport is superb. I have been hunting 'coons for one reason or another ever since.

Any 'coon hunter or trapper will tell you there are many kinds of raccoons. He may recognize cave 'coons, swamp 'coons, hill 'coons, or pigmies, depending on where he comes from. E. Raymond Hall recognized thirty-two different kinds of raccoons in North America (see 1981 reference under Mammals, above). I do not believe there are that many.

Our problems in the Keys began in 1930 when E. W. Nelson described and named four new subspecies in addition to the Florida raccoon, *Procyon lotor eleucus*, already described by Outram Bangs in 1898. Nelson named *auspicatus* from Marathon on Key Vaca, "*inesperatus*" from Upper Matecumbe, "*incautus*" from the Torch Keys west of Big Pine, and *marinus* from Chokoloskee over in the Ten Thousand Islands of Collier County. I regard only the first and last as valid, and I interpret their ranges and characteristics quite differently from Nelson. Major E. A. Goldman (1950) summed up the work of Nelson, his own with Nelson on coastal island forms from Georgia and South Carolina, and the named forms I believe were introduced to the Bahamas and Lesser Antilles. Because of the extreme variability of raccoons in color, size, skull shape, and teeth, the whole thing is extremely complicated.

I got involved because of the raccoons on Guadeloupe and Barbados, in the remote Lesser Antilles, and the pigmy form on the western Caribbean island of Cozumel. I wanted to know if those were really different kinds of raccoons (from *P. l. lotor*) and, if so, how different (species or subspecies). The problems were so thorny my overview took years to develop. I have examined more than three thousand specimens.

At first things seemed simple enough. There were big-toothed raccoons and small-toothed raccoons. The Barbados and Guadeloupe raccoons looked like introduced populations of each of those, respectively. The pigmy raccoon seemed to be a truly distinct, native form on Cozumel: *Procyon pygmaeus*.

Then I got to South Florida. I will not herein regale you with all the tales of nights in the Everglades and journeys to remote Keys. With friends and colleagues, over several years, I hunted and trapped hundreds of 'coons, and approached their classification in every way and with every method I could think of. One morning—quite early—in the air-conditioned research labs at Everglades National Park, Dr. Robert Rosenbaum, a blood expert from Northwestern University, and I were centrifuging and decanting, using the whirring machines and test tubes and clad in the white lab coats of real scientists. Bob looked up and said:

"Nobody would ever believe that an hour ago we were out there, eight miles west of here, in the dark, covered with mud and blood and mosquitoes, both fighting as hard as we could just to hold that big ol' mean mother down. . . ."

What Bob and I were doing is called "blood protein electrophoresis." This sounds like some wonderfully arcane scientific procedure, but it is really very simple. What one does is separate a blood sample into the clear serum and the blood cells, then discard the cells. The serum is then "run" in a medium of known density and a field of electric charge. The protein molecules in the serum will respond to the charge in accordance with their own charges. Molecules with net positive charge will move toward the cathode, or negative pole, of the field. Negatively charged molecules will move toward the anode. The amount of charge and the size and shape of each molecule will determine just how fast it will plow through the medium.

After a set time, identical for all samples being run in the comparison, the field is switched off and the medium is stained to show the positions of the proteins. One then may graph the positions of the proteins relative to each other and the neutral point of origin where each sample was put into the medium.

I herein provide you with the picture Bob and I developed. Our sample sizes were small, but the pattern was most impressive: the only three sorts of raccoons I could recognize on grounds of anatomical characters and coloration did seem to be different in electrophoretic profile. Because a protein represents a direct decoding of a gene, made of DNA, different peaks on the graph correspond to real genetic differences in the animals. The *height* of a given peak just corresponds to the *amount* of that protein present in one individual's blood. It is the *position* of the peak—right or left of the origin—that matters.

This sort of work should be done for dozens more raccoons from all sorts of populations. Based on morphology (size, skull characters) and coloration, I now recognize three sorts of raccoons in South Florida and the keys. For definitions of terms and measurements see Hall, 1981 (cited in Mammals, above). All are big-toothed. Here is a quick synopsis:

1. Florida raccoon, *Procyon lotor eleucus* Bangs (1898). Usually large raccoons with pelage richly suffused with yellow, orange, or reddish, especially on the nape and shoulders. The tail rings are bold and complete. Fur closely approaches the eye, to within 2 mm in the fresh animal. The skull is broad and highly bowed in the frontonasal region. The frontal height, measured normal to the palate just behind the last (second) molars, is as great or greater than the length of the muzzle, measured from just inside the front of the zygomatic arch to the tip of the premaxillaries. This is an upland animal. In South Florida it occurs on hammocks and pine islands in the Everglades and on some of the larger Florida Bay Keys (e.g., Black Betsy Keys).

2. Marine raccoon, *Procyon lotor marinus* Nelson (1930). Usually small raccoons with somber grey pelage and a brown tint to the underfur of the nape. The tail rings are complete; the extent of bare skin around the eye is 2 mm or less. The skull is not especially wide or bowed in profile. The frontal height is less than the muzzle length. (Young specimens are hard to identify. To do it, use a complex ratio: the quotient of frontal height divided by muzzle length is divided into the quotient of condylobasal skull length divided by maximum zygomatic width of the skull. The result is less than 175 in *eleucus*—average *ca* 170—and greater in *marinus*—average *ca* 180.) This is the raccoon of the sawgrass 'glades and mangrove swamps, common on the little Keys of upper Florida Bay.

3. Key raccoon, *Procyon lotor auspicatus* Nelson (1930). Variable in size and pale in coloration. This form varies from grey to almost pinkish, and usually has the pallid pelage suffused with yellow, orange, or pink tones. The tail rings are usually incomplete, especially below. The mask may be weakly indicated too, but the major point of identification is the large, nearly black areolus of bare skin around the eye—at least 3 mm wide. The skull is variable and intermediate between that of the two preceding subspecies. This raccoon occurs from Grassy Key throughout the Middle and Lower Keys to Key West, and out in the Gulf as far as the

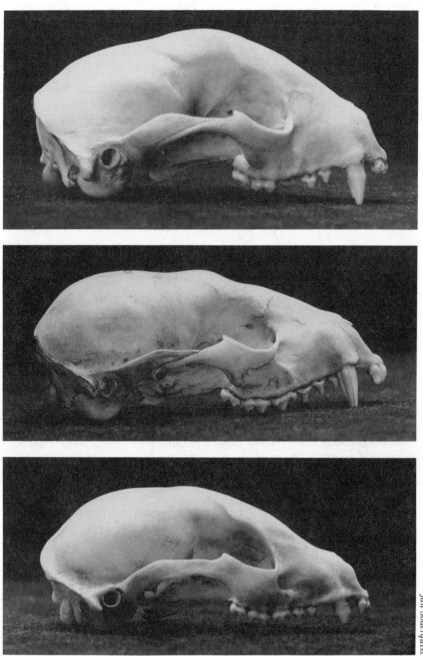

Jan Soderquist

Top: Florida raccoon, *Procyon lotor eleucus,* from Merritt Island.
Middle: Marine raccoon, *Procyon lotor marinus,* from Everglades.
Bottom: Key raccoon, *Procyon lotor auspicatus,* from Boca Chica.

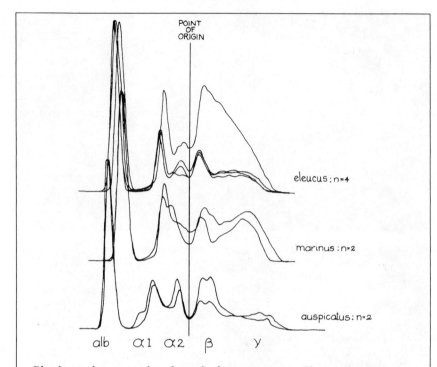

POINT
OF
ORIGIN

eleucus : n=4

marinus : n=2

auspicatus : n=2

alb   α1   α2   β   y

Blood samples were taken from the living raccoons. The results show the electrophoretic mobility of the alpha-1 globulin to be different for the *auspicatus* individuals compared to the others. The *marinus* individuals can be most readily identified by the presence of a distinct gamma peak. The *eleucus* individuals are similar to each other and differ from the others in the lack of a specific gamma globulin peak. The point of origin marks the point of electrical neutrality for each specimen. The anode is to the left, the cathode is to the right. Five peaks can be identified: 1) Albumin (alb); 2) Alpha-1 globulin (α 1); 3) Alpha-2 globulin (α 2); 4) Beta globulin (ß); and 5) Gamma globulin (Y).

Content Keys. Its kidneys, however, are no more modified than mainland raccoons' and it cannot drink sea water (Dunson and Lazell, 1982).

Now you will note that I have not named the raccoons from the Upper Keys. That is certainly not for lack of 'coons there. Nor is it for lack of an available name: remember old *"inesperatus"* of Nelson (1930). I believe the Upper Keys raccoons represent an unnameable hodgepodge and blend of all three of the recognizable subspecies. You might find an individual that identifies as any one of the three, or any combination of characters in between. Many Key Largo raccoons fit *eleucus* pretty well; several Long Key raccoons fit *auspicatus*. But do not be surprised by the numerous exceptions.

Dr. Michael Bogan of the Denver Wildlife Research Center, U.S. Fish and Wildlife Service, also examined many specimens of Keys raccoons. His results, reported to the American Society of Mammalogists orally, disagree with mine. He believed all of Nelson's Keys' subspecies could be recognized. I tried to repeat his measurements, using exactly the same skulls, and the same program. I got completely contradictory results. I believe the name "*inesperatus*" belongs to a three-way intergradient set of populations, and that "*incautus*" is a flat-out synonym for the raccoon first named *auspicatus* at Marathon.

If I had known then what I know now about the great geographic range of *P. l. auspicatus,* and its relative abundance today right in Marathon, I would never have claimed it to be "threatened" as I did in an account written about 1974 and published in 1978. My misbegotten notion was based on the erroneous belief that *auspicatus* was confined to the Middle Keys and scarce on Key Vaca. Neither is remotely true. These egregiously abundant garbage can raiders are not at all threatened.

How did this complex assortment of raccoons in South Florida evolve? In an attempt to answer this question I examined the considerable fossil material housed at Florida State Museum. First, the simplest of the three forms:

The Key raccoon first reached what are today the Middle and Lower Keys during the Wurm glacial maximum. The ancestral stock was either *eleucus* or *marinus* or an intergrade between the two; it hardly matters or is discernible. *P. l. auspicatus* evolved after the Wurm, as rising sea level isolated the present Keys and they became increasingly saline and arid. The pallid coloration is an adaptation to high heat and insolation: it is reflective. The bare skin around the eye is a bit mysterious, but may result from a general lessening of overall pelage density—also a selected response to heat. Genetic continuity with mainland—or at least Upper Keys—raccoons has never been totally lost, so a blend of skull and size characters has been either maintained or achieved subsequent to sea level rise. Key raccoons have long tails, but the difference between them and their more proximate relatives is weakly modal. I believe their long tails are a heat-dissipating mechanism. This is discussed under Key cotton rat—another relatively long-tailed form—below.

During the Sangamon Interglacial, and during preceding Interglacials of the more distant past, South Florida was inundated. On the Highlands Island above Okeechobee the *eleucus* raccoon evolved in isolation. It almost evolved all the way to full species status, but not quite. Raccoons intermediate between *eleucus* and the *marinus* raccoon of the (remote) mainland coast occupied the scattered islands of what is now central Florida. As sea level dropped with the onset of the Wurm, raccoons of both sorts—and their intermediates—invaded new land rapidly. The marine raccoon seems to have had the competitive advantage in swamps and

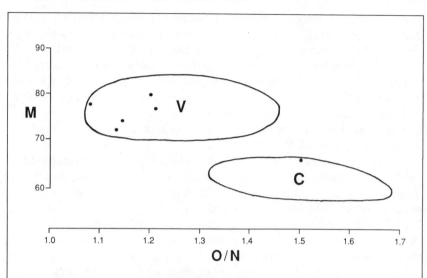

Morphometrics sound arcane and difficult, but when species are very distinct they are just simple arithmetic. Using the three measurements shown in the diagram of a Key deer skull, one can readily separate all Key deer, within range C (for *clavia*), from all white-tails, within range V (for *virginiana*). I plotted the length of the molariform row, M, against the ratio orbitofrontal width, O, divided by nasal length, N. The dot within *Dama clavia* range is the type-specimen, old buck MCZ 19120; he has the biggest teeth measured for his species: 66.4 mm. The dots within *Dama virginiana* range are for South Florida specimens of the subspecies *seminola*. Their position indicates character divergence from *D. clavia*. The evolutionary phenomenon of character divergence cannot occur between populations of the same species because it results from selection against genetic continuity. My data are from 28 *D. clavia* and 100 *D. virginiana* at YPM, Yale, MCZ, Harvard, and NMNH, Smithsonian.

marshes, but *eleucus* held the high ground. During the Wurm maximum marine raccoons were far out at the edge of the continental shelf; all of today's Florida mainland was occupied by *eleucus*. But as sea level began to rise about twelve thousand years ago, the situation reversed itself.

In the Everglades and upper Bay Keys, for example, *eleucus* still holds the high ground. Here *eleucus* and *marinus* act like full species: they are well niche-segregated by habitat; big *eleucus* tend to be groupy, troopy fellows and respond immediately to the alarm squall of other raccoons. On the other hand, little *marinus* 'coons tend to be solitary and bolt from alarm squalls. Intermediate raccoons did well around the inland reaches of Tampa Bay and Appalachicola. These are vast zones of intergradation today.

On what became the Upper Keys the pattern looks more like introgressive hybridization. As sea level rose the areas of distinct upland versus marsh habitat decreased. Soon, perhaps, there was too little of each to support discrete populations; lack of mate choice selection resulted in the mix of phenotypes we see today. In the upper Bay, stranded populations have maintained themselves, but excluded whichever sort had numerical inferiority or the lesser habitat.

Blood proteins support this scenario: *auspicatus* has a profile roughly intermediate between the very divergent *eleucus* and *marinus* types.

The marine raccoon is widespread today. It occupies the Gulf Coast westward through Louisiana (at least) and the Atlantic coast north to Cape Romain in South Carolina (at least). It occurs in a wide band across southern Georgia in the Okefenokee region. I have examined hundreds of specimens from this inland portion of the range, corresponding to the old Sangamon mainland coastline, especially at the American Museum (AMNH).

A population biology study of the marine raccoon was done by Bigler and colleagues (1981). They found this subspecies has a broad reproductive season, with females ready to mate in April and births occurring through October. This contrasts with the sharply defined breeding season of more northerly populations, which produce young in May and June.

The reproductive rate is low, with litters averaging less than three young. Only about one half of the population survived for three years or more. These raccoons could cross seawater gaps of up to 645 meters. During the four years of their study, Bigler et al. found population density varied from 2.2 per square kilometer (in May) to 6.5 (in December). Most of that difference is probably young of the year who will not survive their first dry season (winter).

I know of no comparable study for our endemic Key raccoon, *P. l. auspicatus*. I would expect both similarities and differences. It is amazing to me that such abundant, conspicuous, easily captured, and highly popular animals remain so little known. Where are the naturalists of yesteryear?

References

Bangs, O. 1898. The land mammals of peninsular Florida and the coast region of Georgia. *Proceedings of the Boston Society of Natural History* 28: 157–235.
Bigler, W., G. Hoff, and A. S. Johnson. 1981. Population characteristics of *Procyon lotor marinus* in estuarine mangrove swamps of southern Florida. *Florida Scientist* 44(3): 151–57.

Dunson, W., and J. Lazell. 1982. Urinary concentrating capacity of *Rattus rattus* and other mammals from Lower Florida Keys. *Comparative Biochemistry and Physiology* 71A: 17–21.

Goldman, E. A. 1950. Raccoons of North and Middle America. *North American Fauna* 60, Washington, D.C.: U.S. Fish and Wildlife Service.

Lazell, J. 1972. Raccoon relatives. *Man and Nature,* Massachusetts Audubon Society, September: 11–15.

———. 1973. Ecology and taxonomy of raccoons. *Annual Report,* Office of the Chief Scientist, National Park Service, USDI: 82–83.

———. 1978. Threatened Lower Keys raccoon. In *Rare and Endangered Biota of Florida* 1: *Mammals,* J. Layne, ed., pp. 25–26. Gainesville: University of Florida Presses.

———. 1981. Field and taxonomic studies of tropical American raccoons. *National Geographic Society Research Reports* 13: 381–85.

Martin, R. A. 1974. Fossil mammals from the Coleman II A fauna, Sumter County. In *Pleistocene Mammals of Florida,* S. Webb, ed., Gainesville: University of Florida Presses, pp. 35–99.

Nelson, E. W. 1930. Four new raccoons from the keys of southern Florida. *Smithsonian Miscellaneous Collection* 82: 1–17.

KEY RABBIT
(SYLVILAGUS PALUSTRIS HEFNERI)

With the advent of another war, the Naval Station underwent another vigorous reactivation. From 1941 to 1945 it expanded tremendously, . . . and assumed jurisdiction of Boca Chica . . . an island eight miles northeast.

SHERILL AND AIELLO, *KEY WEST* (1978)

Earle Rosenbury Greene was an avid naturalist. . . . He was born in Atlanta, Georgia, in 1886. . . . In 1918 he entered the Army as a Second Lieutenant. . . . He went with the U.S. Fish and Wildlife Service in 1934 (then known as the Biological Survey). Later he was with the office of the Inspector of Shipbuilding for the Navy Department. It was in this capacity . . . that he was at the Key West Naval Air Station, Boca Chica Key, Florida.

HARRIETT G. DIGIOIA (*IN LITT.,* 1 APRIL 1985)

From February 1939 until October 1942, Earle Greene was refuge manager for both Key West and Great White Heron National Wildlife refuges.

Always primarily an ornithologist (in which context you may read far more about him and his work in the Keys, below), his general interest in wildlife was pervasive. Much of his work was perforce on the backcountry Keys and mangroves that were included in the refuges, but during 1942 he spent a lot of time on Boca Chica.

In February he recorded red-backed sandpipers there; in April his latest record for sparrow hawks; in May, marsh hawk; in June, western sandpipers and sanderlings; and in July, nesting red-winged blackbirds. But on the eleventh of August 1942, just two months before he left the Keys, Earle secured his greatest prizes.

In June 1941, Roger Tory Peterson first called Earle's attention to the call notes of Lower Keys nighthawks. Peterson suspected they were Antillean birds. Earle does not say why he let a year slip by before he collected specimens; perhaps he needed a secluded spot—as opposed to Key West—for safe shooting. He does note the rapid habitat destruction by spring and summer of 1942: ". . . war activities, including dredging, bombing, and target practice in this area." Perhaps that spurred him to collect while he still could.

On the eleventh Earle shot a pair of nighthawks on Boca Chica, and, at 7:30 P.M., for good measure, a rabbit. He sent the skin and skull of the rabbit to Major E. A. Goldman, prominent mammalogist at the Smithsonian. He wrote Goldman (12 August 1942) that rabbits were numerous on Boca Chica and others of the Lower Keys. Major Goldman wrote back (8 September 1942) that the rabbit apparently was new and represented an undescribed form, but that additional specimens were needed to prove the issue. Earle did not go back and get more. I did.

Forty years later, almost to the day, I looked up at the two young naval officers chatting on the verge of the runway. The officers were paying no attention to me, the peculiar biologist they had been charged with escorting around the naval air station. I knelt, pulled a branch down in a deep bow over the run between the prop roots, through the Key grass. When I let it go, the whistling whip was perfect.

In my mind I was far from mangroves and Key grass, back in hedgerows of Pennsylvania farmlands. My fingers found the nylon in my pocket and formed the knot quickly: a knot that only tightens, never loosens. Back in Pennsylvania it had been made in butcher's cord; nylon hadn't been invented then. I could put four neckers along here in twenty minutes, while the officers mused over anyone so odd as to count the little spheres in the piles of rabbit scat. I had asked them if they would shoot a rabbit for me: No. Could I trap one with a box trap myself—alive, unharmed? No. They were extremely cooperative and friendly, but militarily firm: no wildlife of any sort could be taken on Boca Chica. I could look, but not collect.

My situation was desperate. I had several fine specimens of "my" rabbit.

They were very different from all other rabbits known to science; I was certain they represented a new, undescribed form. But all of mine were from one small area—less than three hectares—on Sugarloaf. I needed specimens from other localities. I needed more specimens, period. Local folks had begun to complain about my depredations on the little Sugarloaf colony. But I had a species to describe and name.

A note was in my mailbox when I got home: call Lois Kitching. That would mean something good! She is an astute local naturalist.

"I've got a dead bunny for you," she said. "He's in beautiful shape, must have just been hit. He's from Geiger Key. . . ."

A few weeks later I added a specimen from a third locality. I did it by pulling out a drawer of unidentified rabbit specimens at the Smithsonian. There was Earle Greene's specimen: "Marsh Rabbit?? Boca Chica Key."

The currents of life can be remarkable. Not only do they merge and diverge, they often weave to make bizarre patterns. The Key bunny, the Navy, Earle, and I were meantime joining other notable institutions and characters to blend into a tapestry unique in the annals of biological science.

Just before my captures of the Sugarloaf rabbits—the first actual specimens of adults from the Lower Keys I knew about—I had left the oldest of the several Audubon societies (see *Science* 1979 [203]: 314–16).

Seeking other gainful occupation, I met a woman named Phebe Wray who presided over the Center for Action on Endangered Species. Ms. Wray approached me with what seemed an excellent idea: auction off the name of the new rabbit to get the dollars needed to support my research. The practice of giving patronyms to species in exchange for support is an ancient one and examples are legion. Just look, for example, at the scientific names of many Hawaiian birds. I left the matter entirely in Ms. Wray's hands.

She was especially hopeful about two corporations: Volkswagen and Playboy—because of their rabbit-oriented proclivities. In the event, the Playboy Corporation won the bid. They wanted the rabbit named for their founder, Hugh Hefner. I knew there would be problems and some colleagues would object. Through no fault of the Playboy Corporation, a little of the money—far less than one half—actually came through and was spent on wildlife conservation research.

The Key bunny is a somber, button-tailed form quite like the mainland marsh rabbit (*S. p. paludicola*). It differs—the way most sorts of rabbits differ from each other—in skull characters. The Key form has an extremely long dentary symphysis (the joining of the two lower jaw bones at the chin) and a very short molar tooth row. The facial profile is convex and bowed, and the braincase is broad. Forming ratios of symphysis length, frontal height, and braincase width—all relatively large measurements in the Key bunny—with molar row—a short measurement—will separate specimens from any ordinary marsh rabbit.

Two other sorts of rabbits occur—or have occurred—in the Keys. European rabbits (sometimes called "hares," as in "Belgian hare") are common domestic and feral animals. Populations are established on West Summerland, Ramrod, and at both Perkey and on the south coast of Lower Sugarloaf. At the two latter localities the European rabbits, *Sylvilagus* ("*Oryctolagus*") *cunniculus*, and native Key bunnies can occasionally be seen together. They do not hybridize; at least they never have yet. However, the introduced European animals no doubt compete with the native endemic for food.

The second interloper is a true hare, genus *Lepus*. True hares differ from rabbits, genus *Sylvilagus*, in appearing to lack an interparietal bone. This is a prominent, medially located bone in the back of the skull of rabbits. In hares it fuses into the surrounding bones early in embryology. Hares have gestation periods a week or more longer than rabbits and produce precocial young: born furred and with open eyes. Rabbits produce altricial young: born naked and with their eyes closed.

A single specimen of a hare, genus *Lepus* but not yet identified to species, is known from Upper Sugarloaf: YPM 08926. When I first saw this skull, I mistook it for a European rabbit simply because it was big and had prominent supraorbital bones. I didn't think to check for an interparietal. California jacks, *Lepus californicus*, have been introduced into South Florida. I used to see them frequently at Miami airport. Perhaps some got transported down the Keys. I do not know of a living population today.

European rabbits and hares are larger than the Key bunny, have much longer ears, and big, fluffy tails. Many European rabbits are white or particolored. Hares have white bellies and white undersides of the tail—which they hold up. Key bunnies do not display their tiny, all-grey tails.

The Key bunny is certainly endangered today. Landfill and outright predation by human hunters have wiped out whole island populations, as in the Torch Keys. The species is definitely known from Big Pine, Sugarloaf, the Saddlebunch, Geiger Key, and Boca Chica. Scats have been reported from other Keys, but deer scats can be confusingly similar, and I am reluctant to chronicle records without actual specimens. Rabbits are extremely fecund, as everyone knows. Removing one voucher specimen will not endanger a population on a back-country Key. The Fish and Wildlife Service should obtain voucher specimens for all Keys under its jurisdiction that have rabbits.

There is evidence of geographic variation in Key bunnies. The few available Big Pine skins are more richly colored dorsally, and have whiter undersides of the thighs, than those from Sugarloaf to Boca Chica. Typical

S. p. hefneri, from the western part of the range, have grey thighs. With more specimens, it might be possible to test the hypotheses that *hefneri* is a species distinct from *palustris* and that it divides into two subspecies.

References

Dunson, W., and J. Lazell. 1982. Urinary concentrating capacity of *Rattus rattus* and other mammals from the Lower Florida Keys. *Comparative Biochemistry and Physiology* 71A: 17–21.

Lazell, J. D. 1984. A new marsh rabbit (*Sylvilagus palustris*) from Florida's Lower Keys. *Journal of Mammalogy* 65(1): 26–33.

KEY COTTON MOUSE
(Peromyscus gossypinus allapaticola)

The source of the subspecific name is "allapattah", the Seminole word for the dense jungle-like stands of hardwood trees, which are called hammocks, in southern Florida. Since the insular cotton mice occur in greatest abundance in the dense hammock at the northern end of Key Largo, this seems a fitting designation for this brightly-colored form.

Albert Schwartz, 1952

The type-locality for this form is often given as Rock Harbor, but it is really "12 miles northeast of Rock Harbor." That is Upper Key Largo. Dr. Schwartz mentions two specimens from "Planter, Key Largo" collected in 1895 (and now in the NMNH). "Planter" is said to be an old name for what is now part of Tavernier. These sprightly creatures have never been found anywhere nearly that far southwest since.

Along with the woodrats, some of these were transplanted to Lignumvitae Key in 1970. A single stuffed specimen exists—in a drawer—on Lignumvitae. It was caught in a boat, however, not technically on the Key. There is no present evidence to indicate that this transplant effort—so demonstrably successful for the woodrat—worked at all. No specimens have been collected—or even reported seen—since shortly after the transplant attempt.

Everyone seems to associate this species with the Key woodrat, which—vastly larger and more flamboyant—has attracted far more specific attention. Both species certainly occur in the hardwood hammocks of North Key Largo, and the cotton mouse seems to even use the woodrats' stick nests occasionally. Recent work by Numi Goodyear, however, indicates that cotton mice often occur in burned-over habitats marginal for woodrats. In these situations the mouse is usually associated with the red-rumped cotton rat, discussed below.

Why all this "cotton"? The specific epithet *gossypinus* means "of cotton," but whether made of or living in is not clear. Cotton mice do not frequent fields or other agricultural areas. The advent of cotton farming in the Southeast no doubt seriously diminished the habitat of this species. Perhaps the animal was named for what the original specimens were stuffed with.

Cotton rats, see below, are associated with fields and farms, so their name may be more appropriate.

No one has ever questioned the validity of the Key cotton mouse as a distinct form. The original description is in rather cavalier form, conspicuously lacking the detail, quantification, and comparisons found in Sherman's description of the woodrat. Dr. Schwartz says *P. g. allapaticola* "exceeds" the mainland races in all sorts of measurements: total length, tail, ear, greatest length of skull, and upper tooth row. However, just averaging larger may not be good enough to separate many specimens. One wants to know the extent of overlap.

Fortunately, the nasal bones apparently average *shorter* in the Key form than in those from the mainland. Predictably, a ratio, say nasal length as percent of maxillary tooth row, would separate all or most specimens. Dr. Schwartz really based his description on color characters, as my quotation suggests. Our Key cotton mouse is reddish on the back with comparatively grizzled sides. The mainland forms are pinkish to yellowish with less grizzled sides and a middorsal zone more heavily overlaid with near-black guard hairs. These are the sorts of characters one needs to see in direct comparison to appreciate.

Cotton mice are closely related to the white-footed and deer mice of more northern climes. They are endearing little sprites of the forest, with boldly contrasting light underparts and feet, and lightly furred, white-bottomed tails. They certainly deserve more study.

References

Brown, L., and R. Williams. 1978. The Key Largo woodrat (*Neotoma floridana smalli*) and cotton mouse (*Peromyscus gossypinus allapaticola*) on Lignum Vitae Key, Florida. *Florida Naturalist* 44: 95–96.

Goodyear, N. S. 1985. Results of a study of Key Largo woodrats and cotton mice. Phase I, spring and summer, 1985. M.S., Zoology, University of Rhode Island, Kingston, R.I. 02881.

Humphrey, S.R. 1988. Density estimates of the endangered Key Largo woodrat and cotton mouse (*Neotoma floridana smalli* and *Peromyscus gossypinus allapaticola*), using the nested grid method. *Journal of Mammalogy* 69 (3): 524–31.

Schwartz, A. 1952. Three new mammals from southern Florida. *Journal of Mammalogy* 33(3): 381–83.

KEY COTTON RAT

(SIGMODON HISPIDUS EXSPUTUS)

In April, 1920, Mr. Winthrop Sprague Brooks collected two adult cotton rats . . . on Big Pine Key . . . which are so different from those of the neighboring mainland, that they seem worthy of recognition as representatives of a distinct island race.

GLOVER MORRILL ALLEN, 1920

Dr. Layne, in his major paper cited above in Mammals, says: "The cotton rat is by far the most ubiquitous mammal in South Florida." Actually, there are said to be two distinct forms in the Keys. The endemic form *exsputus*—whose name means "cast out, banished, exiled"—is found on the Lower Keys, beyond the Seven Mile Bridge. On Key Largo occurs the red-rumped or Cape Sable cotton rat, *S. h. spadicipygus* Bangs (1898).

Both forms are said to be smaller than the subspecies found farther north on the Florida peninsula. As its name implies, the red-rumped form has its posterior "cinnamon-rufous," whereas the Lower Keys form has a general dorsal wash of a "pale ochraceous tone." Both are salt-and-pepper, grizzled creatures with long black guard hairs. The belly of the red-rumped is grey with nearly black-based hairs. In *exsputus* the belly is nearly white with slate-grey hair bases.

The most important difference is tail length. The tail is 35 to 41 (average 37) percent of total length in the red-rumped, but notably longer—43 to 48 (average 45) percent in the Lower Keys form.

I ascribe to the view of the Smithsonian's Richard Thorington that a longer tail is beneficial in a hotter climate. He points out that a rat "uses its tail to dissipate heat. The tail is vasodilated and the temperature of the tail rises above that of the environment. This reaction is important in the

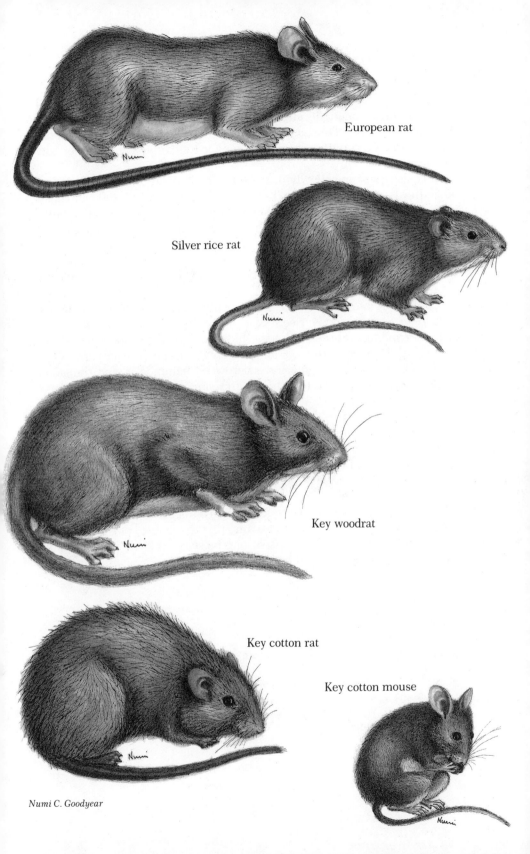

European rat

Silver rice rat

Key woodrat

Key cotton rat

Key cotton mouse

Numi C. Goodyear

thermoregulation of these animals and it appears to have an important bearing on the growth of the tail." He presents ample evidence that tail length is environmentally affected in various kinds of rats (but not specifically in *Sigmodon*). Thus, it is at least questionable whether the longer tail of *exsputus* is a genuinely heritable, genetic trait, or just environmentally induced.

It is also questionable whether the Lower Keys are significantly hotter than Key Largo or Cape Sable. True, meteorological records indicate that it has never gotten as cold in the Lower Keys as it has farther north, but it may also never get as hot either. The sea has an insulating effect, moderating temperature extremes.

No one knows what sort of cotton rats occupy the Keys from Plantation to Key Vaca. There are cotton rats on them, but no one has critically examined them.

Cotton rats are blunt-faced, hamster-like rodents with much shorter tails than European rats, rice rats, or woodrats. Their teeth are the flat-topped grinding sort seen in horses and woodrats, not cusped as in European rats. They are terrestrial and usually live in burrows or runways in dense ground cover in pinelands, hardwood hammocks, and transition zone. They often get into gardens and yards—where they can cause considerable damage—but they rarely enter houses.

The Key cotton rat, *S. h. exsputus*, is uncommon except in back-country areas away from human habitation. Its populations may well boom and bust, however, and at times it can be abundant. With a little land acquisition—especially of fresh water sources—the threat to this form's survival would be removed. Like most small rodents, these animals are beneficial insect eaters and major consumers of vegetable matter. They are a crucial link in the ecosystem supplying food for predatory birds, other mammals, and snakes. I think it is safe to assume that the demise of the cotton rat would seriously endanger some of our most remarkable and desirable species, like the short-tailed hawk.

References

Allen, G. M. 1920. An insular race of cotton rat from the Florida Keys. *Journal of Mammalogy* 1: 235–36.

Bangs, O. 1898. The land mammals of peninsular Florida and the coast region of Georgia. *Proceedings of the Boston Society of Natural History* 28: 157–235.

Thorington, R. W. 1966. *The Biology of Rodent Tails a Study of Form and Function*. Fort Wainwright, Alaska: Arctic Aeromedical Laboratory.

KEY WOODRAT

(NEOTOMA FLORIDANA SMALLII)

Each day the animal kingdom forced its presence on us. There we
walked into a rat town. The large wood-rats of the island select a
tree from beneath which they can excavate the humus. This ... rat
usually builds a shack two to three feet wide and four to six feet
long, composed of sticks one to three inches in diameter. These are
piled up to a height of one to three feet in an irregular fashion.
Beneath this structure they carry on their domestic operations.

JOHN KUNKEL SMALL, 1923

Thus it was that the botanist Dr. Small became the first scientist to publish on the Key woodrat in the hardwood hammocks of Key Largo. It would be a quarter of a century—John Small dead and gone, but not forgotten—before Thomas Barbour, of Harvard, called the attention of Harley Sherman to Small's words. Sherman and Joseph C. Moore set out posthaste from Gainesville and captured six specimens. Sherman located and borrowed more, including twenty-six—more than half his eventual total of forty-six—from Albert Schwartz of cotton mouse renown.

Sherman did a meticulous and detailed job of comparing the Key Largo animals with their relatives. There were no relatives—native stick-nesting or "pack" rats of the genus *Neotoma*—in South Florida. The species, despite its name, is a north-temperate climate form; the Appalachians are its real headquarters. Today on the Florida peninsula it penetrates as far south as Englewood on the Gulf Coast and Vero Beach on the Atlantic. A subfossil from a solution hole near Princeton, in Dade County, indicates the species was recently present much farther south. So far as I know, that subfossil skull has not been critically examined to see if it is nominate *floridana* or *smallii*—or an intermediate.

Sherman's description, thorough as it is, is tediously difficult to follow. Its accompanying drawings do not clearly indicate the differences between a Key woodrat and an ordinary one. I have examined some (not all) specimens, and believe the magnitude of the distinctions to be greater than Sherman perceived. Most readers are not too interested in details of cranial osteology, so I relegate my findings to an illustration. Most of you want to know just what is so wonderful about this rat anyway.

The animal kingdom has forced its presence on the land developers and financiers who would convert Upper Key Largo. They have walked into a rat town indeed. There are those of us who will do anything legally possible to prevent the destruction of the Key woodrat and its habitat.

This is a big rat, to over 16 inches (420 mm); the tail is about 45 percent of the total length. The ears are very large—over an inch (26 mm) high. The eyes are large and prominent. The fur is short but soft; fine fur extends to the tip of the tail (the tail is not ringed with scales like a European rat's). Dorsally the woodrat is warm brown; below cream to nearly pure white. Its molar teeth are flat-topped grinders—like horse teeth—not cusped like the molars of *Rattus*.

This is an agile climber, spending much of its nocturnal activity in the canopy searching for nuts, fruits, and insects to eat. Of course that is true of *Rattus rattus* too, and it is also big and sometimes similarly colored. Alas, even fewer people can tell a native woodrat from a European sewer rat than can tell a Cape Sable seaside sparrow from an English sparrow, or a white-crowned pigeon from a European street pigeon. More's the pity.

Eastern woodrats are rather docile and have a cuddly, teddy-bear quality. They avoid humans wherever possible and rarely enter human dwellings. Unfortunately, they seem to have a hard time coexisting with humans, at least on Key Largo. Cats kill them, dogs dig out their homes, and European rats—garbage eaters—compete with them wherever hu-

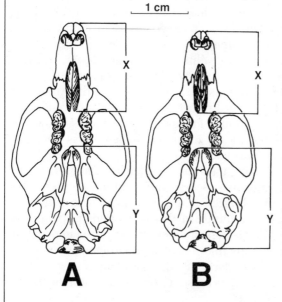

How to identify Florida woodrats. X is the distance from the tip of the nasals to the front of the palatal bridge. Y is the distance from the back of palatal bridge to the condyles. Both average greater in the Key woodrat. The datum X is transformed by subtracting it from a constant. Thus, 30-x/Y separates the forms. Redrawn from Sherman (1955).

mans live. *Homo sapiens* generates vermin, and woodrats are elegant creatures quite unaccustomed to that.

The habit of home construction is shared with the beaver, the rare and endangered stick-nest rats of Australia (genus *Leporillus*), humans, and few other mammals. Key woodrats are not required to live in stick nests. As Small pointed out, holes and tree stumps will do. However, wherever possible these woodrats seem to like to erect an edifice. I know little of the internal design of their homes except that there is said to be at least one chamber and at least two entrance-exits. But about interior decorating I do know a bit. This is the most human-like trait of the Key woodrat.

Many sorts of woodrats are avid collectors; that is how some of the western ones got the name "pack" rat: they store things. Our woodrats tend to like things that are either very pale (white is fine) or very brightly colored. They love shells, just like humans, but have the good sense to collect empties rather than kill off the population of living snails. When woodrats were first established on Lignumvitae Key it was feared they might consume some of the gorgeous tree snails that live there. They do collect the shells, and decorate their homes with them, just like us. They do not seem to depredate the snail populations.

They also collect skulls. This seemingly macabre habit is shared with humans too, especially me. My home is decorated with the heads of dead animals. Woodrats do not have access to taxidermists, so they cannot display mounted heads. However, I maintain about 90 percent of my collection as skulls too; scientifically these are more valuable than mounted heads. Woodrats do not currently utilize fish. This is simply because mounted tarpon and sailfish are too big to fit over their mantelpieces. If someone would mount killifish or Key silversides (*Menidia conchorum*, another rare and endangered species) I know the woodrats would queue up to get them.

Quite unlike me, but like a large number of other humans I know, they collect rings. I am sure that gold and diamond rings would enthrall them, but Upper Key Largo provides few of these. Instead Key woodrats love the aluminum rings of some kinds of pop-tops and the brilliant rings from the tops of plastic milk bottles (blue is no-fat, yellow low-fat, and red full-strength).

An abandoned woodrat house may be a veritable treasure trove. One may find gaudy snakes—coral snakes, or red or yellow ratsnakes; or a fine, big diamondback rattler; or exquisite rarities like the rimrock crowned snake. One may obtain excellent skulls of lizards, birds, or the little-known Key cotton mouse. I have no use for plastic rings, but I do hope someday a Key woodrat will have turned the tables on *Homo sapiens* and exploited them by sequestering a hoard of metallic and stone rings snitched from the occupants of Ocean Reef.

There is a place for vacationing tourists in this world and it is called

Miami Beach. There is also a place for the Key woodrat. Let's leave the woodrats in their homes in the hammocks in peace.

The Key woodrat is today the most distantly disjunct mammal living in the Florida Keys. All others have close relatives much nearer. Should the distinctions between this form and its mainland relatives prove absolute, I would urge that it be granted full species-level status. The best place to see the homes of these animals—and maybe an individual itself—is on Lignumvitae Key, to which the species was introduced (see the Brown and Williams reference under cotton mouse).

References

Hirschfeld, S. E. 1968. Vertebrate fauna of Nichols' Hammock, a natural trap. *Quarterly Journal of the Florida Academy of Scientists* 31: 177–89.

Humphrey, S.R. 1988. Density estimates of the endangered Key Largo woodrat and cotton mouse (*Neotoma floridana smalli* and *Peromyscus gossypinus allapaticola*), using the nested grid method. *Journal of Mammalogy* 69 (3): 524–31.

Sherman, H. B. 1955. Description of a new race of woodrat from Key Largo, Florida. *Journal of Mammalogy* 36: 113–20.

Small, J. K. 1923. Green deserts and dead gardens. *Journal of the New York Botanical Garden* 24(286): 193–247.

SILVER RICE RAT
(ORYZOMYS ARGENTATUS)

The life of the silver rice rat trapper is like that ascribed to the Maytag Repairman. Our motto: don't hold your breath.

How can a male animal that covers so much territory and sleeps in so many different nests be rare?

NUMI GOODYEAR

On the last day of January 1973, we decided to go to the Keys. "We" included a half-dozen of my biology and ecology students from a small prep school where I taught at the time. One of those students was Numi Catherine Spitzer—now Goodyear. We had been in the Everglades searching out—among other things—rice rats and kingsnakes. These two species enjoy a remarkable ecological and ethological, coevolutionary situation on the Outer Banks of North Carolina. Someone had suggested that Florida kingsnakes were closely related to the endemic Outer Banks form, and we had decided to see if the special relationship existed down in the 'glades. It did not, but we were having excellent luck anyway. Among other things, we had salvaged a road-killed panther.

Anyway, we went all the way to Key West, then retreated back to the Big Pine Fishing Lodge: more our sort of place. On the first of February we met Jack Watson. We spent the next several days exploring Big Pine and the Torch Keys, and strayed as far as Summerland and Key Vaca. On the fifth, one student, Anne, developed a toothache. Anne's toothache needs to be recorded in the annals of science. Because of it we drove her to the dentist in Key West.

On the way, Numi spotted the Cudjoe Key cattail marsh. Numi was the acknowledged expert on rice rats, which are aquatic relatives of woodrats and cotton mice. She had not only discovered them on the Outer Banks, but had figured out their association with the peculiar kingsnake there.

"There are rice rats in that marsh," Numi said.

"Ridiculous," I said. "These Keys were discovered in 1513 and every mammalogist in America has been down here poking around. If there were any rice rats in the Florida Keys they would have found them."

"Well, then they didn't look there," Numi said, "because there *are* rice rats in that marsh."

We had passed the marsh at certainly 50 mph, and it is just a little thing. But on the way back from Key West Numi persuaded me to stop long enough to set a wire mesh live trap. We came back the next morning; I recorded:

"Checked trap: scats, but no mouse; was it *Oryzomys*? right size. . . ."

I was in love with the pinelands on Big Pine. We spent the day surveying land I eventually bought, but Numi insisted on buying five big Victor Four Ways rat traps. She would get more than scats next time. We drove down from Big Pine to Cudjoe between five and six in the evening and she set her snap traps.

Impatient, we drove back down again at ten o'clock that night (past my bedtime, usually). Numi had a lovely big *Sigmodon*, cotton rat, of the endemic form, but no rice rat. *Sigmodon* usually produces bigger scats than *Oryzomys*. If Numi was beginning to doubt herself she never let on. She just waked me up early for trap check the next morning, 7 February 1973. My field notes succinctly say:

"*Oryzomys* new form: pale grey and white. . . ."

I went on with more details on the carcass, which had lost a hind leg to scavenging crabs, and specifics of the locality. Numi skinned and stuffed her new rice rat, now in the Smithsonian. We bought three more snap traps and Numi reset nine traps that night. Next morning:

"*Oryzomys argentatus* . . .—MS name here coined. Perfect specimen: silvery grey . . . white below. . . ." That became the type-specimen of the species.

But then our luck ran out. It would be more than seven years before we saw another silver rice rat. We trapped other cattail marshes, and kept trapping at Cudjoe. My field notes include things like ". . . they sure must be rare . . ." and "The rice rat is clearly rare." And it is.

On 9 February 1973, we took Nick Howell, now a botanist at Missouri State University, to the site so he could give us a good description of the vegetation. On 20 February we reached the Archbold Station at Lake Placid, Florida, where Jim Layne had a series of mainland rice rats for Numi to compare with hers. The distinctions—not just in color, but in the skull—were spectacular.

When you think you have a new species in hand your real work begins. Most people in positions of authority are highly skeptical. If you are a 16-year-old high school student they are more than skeptical. Fortunately, Jim Layne could see Numi's specimens, and compare them himself.

I was doing a lot of anatomical work on bears and raccoons in those days, so I traveled often to major museums. I took Numi along, which got her into the inner sancti of the collections, and she measured rice rats. One curator suggested that her "new" rice rat was probably just an introduction from the Antilles or South America. Because genus *Oryzomys* is so widespread and diverse in tropical America, that possibility had to be considered. In fact, Numi did find one with a similar skull: *Oryzomys bombycinus* from Panama. But it is black.

In 1974 Numi went off to university and I went off to Mississippi. Neither of us got to return to the Keys long enough to look for rice rats until 1980. That year I led an Earthwatch expedition. We were busy with Key bunnies, Key ringneck snakes, and other wonders, but we kept the silver rice rat in mind. I had been in the Keys briefly in 1979 and had found what I believed were rice rat tracks on Middle Torch Key. Art Weiner photographed them, and later comparisons have proved me right. You can understand why nobody believed me at the time.

With Dr. William Dunson of Penn State, we were keeping Keys animals in captivity to test their water retention abilities. Bill wanted desperately to at least examine anatomically the kidneys of a silver rice rat, but—sadly— we had not preserved the soft tissues of the first (and only) two specimens. A friend of Bill's, Larry Burns, was having rat problems around his place and was trapping them with snap traps, as would anyone else. Larry said he occasionally got very "pretty" chinchilla grey or silver rats that did not look like the ordinary scaly-tailed ones (*Rattus rattus* of European origin). Bill said please bring him the next one.

On the last day of the Earthwatch expedition Bill called to say Larry had brought him a dead rat. It surely was silver grey. I headed right over to

Summerland Key and finally got to see my third *Oryzomys argentatus*.

I phoned Numi, Jim Layne, and lots of other people. I was ecstatic because a paper had just been published claiming the silver rice rat was extinct. Also, this was the first proven male (the one with the leg chewed off had also lost its genitals). It was much bigger than the female, which is not so strikingly the case with mainland rice rats.

One fine adjunct of Larry Burns's rat trapping was that Bill Dunson got fresh kidneys. He compared these anatomically with the kidneys of other species with experimentally determined water retention abilities. The conclusion was that the silver rice rat "is not likely to be severely limited by habitat salinity." Numi's experiments with members of her live colony indicate weight loss at 18 parts per thousand (ppt) of salt. Seawater is about 35 ppt, so silver rice rats could potentially survive on a 50 percent dilution of seawater.

On that same 1980 field trip I was given a typescript written by biologists from Western Kentucky University in Bowling Green, Kentucky. They had visited Raccoon Key, north of Sugarloaf, prior to its development as a monkey breeding island for rhesus macaques (*Cercopithecus mulattus*). They said they had caught rice rats there. Numi got right on the track and found that they had saved just one skull. She examined that lonely skull: it was *Oryzomys argentatus*.

So, by March of 1980 there were four known specimens from two Keys: Cudjoe and Raccoon. The species was not extinct: it had been "rediscovered" twice. The big male Larry Burns had caught went to the Florida State Museum in Gainesville at the request of state biologists. Numi—by then a graduate student—got some funding at long last to come down and hunt her own rice rat herself. She found them right away—on Raccoon, Middle Torch (thus vindicating me and my footprints), Summerland, and at least six other Lower Keys. Numi has proven their niche to be rather wide. They seem to do well in salt marsh and transition zones, especially in *Batis* swales, but, I suspect, *only* wherever fresh water is accessible as a permanent lens. They have the largest home ranges yet documented in a mammal so small. Their skull characters set them off very sharply from all other North American *Oryzomys*.

The skull of the silver rice rat is proportionately long and slender compared to that of the mainland *Oryzomys palustris* (which is orange in South Florida). The nasal bones are especially long and slender; their combined width at the midpoint is contained in their length more than 4.6 times. Occasionally mainland individuals have nasal bones as slender. When they do, their relatively broad skulls separate them from *O. argentatus*. All mainland specimens with narrow nasals have the width of the skull contained in its condylobasal length less than 1.8 times; it is greater than 1.8 in the silver rice rat because the skull is so narrow.

Numi collared and radio-tracked three individuals on Big Torch and

Summerland Keys. She located each animal hourly all night and day (day is easy: they sleep—except once when high tide flooded one out of its nest). What she found out was utterly different from anything known about any other small rodent on Earth. They may have home ranges of more than 23 hectares, which is 57 acres. No other little rodent ranges so widely. A single individual may build and utilize as many as fifteen different nests. He doesn't just build 'em and forget 'em, either. He remembers where they are and, as dawn approaches, makes a beeline for the one he wants to sleep in that day. He does not always pick the closest one either. On one occasion, a Summerland silver traveled a kilometer—more than half a mile—to get to the nest he wanted from the one he had sacked out in the day before. Other little rodents just do not do things like that. At least not the ones that have been studied. Numi collared and tracked mainland *O. palustris:* they were homebodies, like normal rodents.

Of course, the silver rice rat is one of only a handful of species for which up-close and intimate studies have been made. Generally, we know almost nothing about the lives and life histories of most of the mammals who share our planet. They lead cryptic lives, active at night, out in dense vegetation. If you want to get to know about them, you have to do what they do and go where they go. Modern technology helps, but it is still an awful lot of work few people are willing to do.

The type locality of the species is jointly owned by the Department of Transportation (Route 1 right-of-way) and a private owner. As wetland, it is protected from filling—at least in theory—by the Army Corps of Engineers and other agencies. This choice bit of marsh is about the biggest, best fresh water system left in the Lower Keys. It is prime habitat for indigo snakes, alligators, ribbon snakes, Key mud turtles, Key cotton rats, and a host of other species of plants and animals that are state or federally listed as rare or threatened. It provides fresh water for Key deer. It should be bought and conserved.

References

Barbour, D. B., and S. R. Humphrey. 1982. Status of the silver rice rat (*Oryzomys argentatus*). *Florida Scientist* 45(2): 112–16.

Dunson, W., and J. Lazell. 1982. Urinary concentrating capacity of *Rattus rattus* and other mammals from the Lower Florida Keys. *Comparative Biochemistry and Physiology* 71A: 17–21.

Goodyear, N. C. 1987. Distribution and habitat of the silver rice rat, *Oryzomys argentatus. Journal of Mammalogy* 68(3): 692–95.

———, and J. Lazell. 1986. Relationships of the silver rice rat *Oryzomys argentatus. Postilla*, 198: 1–7.

Humphrey, S. R., and D. B. Barbour. 1979. Status and habitat of eight kinds of endangered and threatened rodents in Florida. *Special Scientific Report* 2, Office of Ecological Services, Florida State Museum, Gainesville.

Spitzer, N. C. 1973. Rice rat's world. *Man and Nature,* Mass. Audubon Society, March: 24–26.

————. 1978. Endangered Cudjoe Key rice rat. In *Rare and Endangered Biota of Florida 1.* Mammals, J. N. Layne, ed. Gainesville: University of Florida Presses, pp. 7–8.

————. 1983. Aspects of the biology of the silver rice rat, *Oryzomys argentatus.* Master's thesis, Zoology, University of Rhode Island (Kingston, R.I. 02881).

————, and J. D. Lazell. 1978. A new rice rat (genus *Oryzomys*) from Florida's Lower Keys. *Journal of Mammalogy* 59(4): 787–92.

· EUROPEAN RATS
(RATTUS RATTUS, AND RELATIVES)

The rats and mice that now live in sewers and cellars, on wharves and farms in the United States originally did not come from this continent at all. They have spread over the world as stowaways, following man to new lands.

ALVIN AND VIRGINIA B. SILVERSTEIN, 1968

These are the loathsome carriers of fleas and plague, the baby biters, food despoilers, and stink sources the land developers want you to keep permanently confused in your minds with our wonderful native rats. I have heard people get up in public hearings and pour forth their horror stories about rats in their efforts to bring about the destruction of natural habitats in the Keys. For all their relevance, they might as well have told stories about man-eating tigers or rogue elephants.

Three species are widespread and abundant now in the United States. We brought them here, and we support and succor them. They do not usually survive long at any great distance from human edifices. The "Norway" rat (thought originally to be from China) and the house mouse exist as domesticated (often albino) animals and are of great economic importance in medical research and the pet trade. The "black" rat, which comes in a great variety of colors, has not been domesticated. It is the most common species in the Keys and occurs even on remote, back-country mangroves. It has specially modified kidneys and can drink water almost as salty as full-strength seawater and survive. It is thought to have origi-

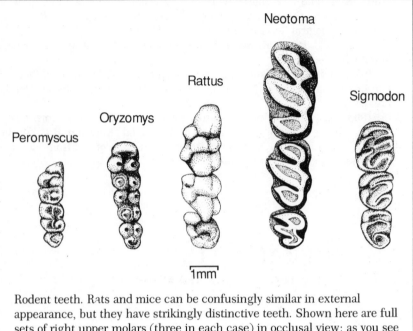

Rodent teeth. Rats and mice can be confusingly similar in external appearance, but they have strikingly distinctive teeth. Shown here are full sets of right upper molars (three in each case) in occlusal view: as you see them with the animal's mouth open wide. Anterior is top. The first three have cusped teeth, like ours. *Peromyscus:* cotton mouse; *Oryzomys:* rice rat; and *Rattus:* European rat (European mouse, *Mus,* has similar, much smaller teeth, with the anteriormost enlarged). The last two have flat teeth, like a horse. *Neotoma:* wood rat; and *Sigmodon:* cotton rat. All are to scale; a one millimeter bar is shown.

nated on the coasts of tropical China, where it occurs in the mangrove forests today.

Most field guides would have you believe the species are easy to identify, but the two rats can be quite confusing. I have delivered to experts specimens that could not be certainly identified. No one, however, has ever presented evidence that they can or will hybridize.

These rats are profoundly different from our native species in tooth structure. Some authorities even put them in a separate family (I won't go that far). They have cusped molars with the cusps arranged longitudinally in three rows. They also have nearly bare, scaly tails, with the scales arranged in rings. Here are the species:

House mouse (*Mus musculus*). Small; not exceeding 250 mm (about ten inches) in total length. The first molar is very large; its length is greater than the combined lengths of the other two molars.

Norway rat (*Rattus norvegicus*). Also called sewer and wharf rat, this is the common species of northern cities, but not very common in the Keys. It is big—more than 250 mm (ten inches) and has a relatively short tail. The tail is said in books to be shorter than head-body length (or less than one half of total length), but this is not always true. The first molar is shorter than the combined lengths of the other two. It has short ears and big feet. The length of the ear is about half the length of the hind foot. This rat is not likely to climb high. It is usually dark brown.

Black rat (*Rattus rattus*). Also known as fruit, roof, palm, alexandrine, or coconut rat. A large species, not as heavy—usually—as *norvegicus*, but very long (to 50 cms or about 20 inches) because of its (usually) very long tail. The first molar is shorter than the combined lengths of the other two. The ears are big—usually more than two-thirds the length of the hind foot. These animals may be nearly black (charcoal grey) above and below, or fawn brown with a white belly, or almost anything in between. They are agile climbers. This is a serious pest species in the Florida Keys, and I would be just as delighted as anyone else to get rid of it. As long as we live the way we do, however, *Rattus rattus* will live with us.

References

Dunson, W. A., and J. Lazell. 1982. Urinary concentrating capacity of *Rattus rattus* and other mammals from the Lower Florida Keys. *Comparative Biochemistry and Physiology* 71A: 17–21.

Silverstein, A., and V. B. Silverstein. 1968. *Rats and Mice, Friends and Foes of Man*. New York: Lothrop, Lee & Shepard.

THE SMALLEST GREY SQUIRREL
(SCIURUS CAROLINENSIS MATECUMBEI)

The very slightly smaller size is merely a reflection of this population's location, at the end of a cline of decreasing size southward through Florida. . . .

HUBBARD AND BANKS, 1970

Here is an admirable problem some interested and enterprising naturalist or biology student might wish to take on. The charming little squirrels of Upper Keys hammocks and backyards were once believed to be a distinct, endemic form. Since 1970, they have been relegated to the synonymy of the South Florida grey squirrel, *Sciurus carolinensis extimus*. But is that correct?

This is the case history: Harold H. Bailey (1878–1962) was an ardent naturalist and son of Harry (*not* Harold) B. Bailey, one of the founders of the American Ornithological Union back in 1883. Harold curated his own Museum and Library of Natural History in Newport News, Virginia, in the early 1900s. By 1922 he was director of the Miami Beach Zoological Park. That seems not to have lasted long, for in 1923 he was president of the Florida Society of Natural History in Miami. During this time he supported himself in the marine insurance and ship brokerage business. He reestablished the Bailey Museum and Library of Natural History in Coral Gables in the thirties, but by the fifties he was running the Rockbridge Alum Springs Biological Laboratory, in Goshen, Virginia. It was in 1928 that he collected Key Largo squirrels.

Harold Bailey coined ten new scientific names for birds and four for mammals. None has fared well. All but one are now regarded as simply junior synonyms for species and subspecies previously named. The sole exception, the Long Island (New York) ruffed grouse, *Bonasa umbellus helmei*, is believed extinct—replaced by introduced stock from the mainland.

Two of Bailey's fourteen names were for the grey squirrel of the Upper Keys. He had to name it twice because his first name had already been used for some other species (no two species can have the same name). Here is the original description:

Sciurus carolinensis minutus. (Sub. sp.)
Little Gray Squirrel

   Type,—B.C.N.H. No. 178, Lower Key Largo, Monroe County, Fla., June 13th, 1928. Collr. H. H. Bailey.
   Smaller than extimus. Lacks the yellowish gray cast of extimus on the back, also the reddish sides of belly. Average length of body,—8.33 inches. Average length of tail,—7.91 inches. Average total length,—16.34 inches.
   This is an island form, found only on some of the lower Keys, off the mainland of Dade and Monroe Counties, Florida.

                                             Harold H. Bailey

This was published as the fourth page of Bulletin No. 12 of the Bailey Museum and Library of Natural History. There is nothing about copyright on it at all. Note that Harold's arithmetic is just about as good as mine. I cheated and used a pocket calculator to get 16.24 inches.

On being informed that his name *minutus* was preoccupied, Bailey renamed the species *S. c. matecumbei*, "the Key gray squirrel," still in 1937. Hubbard and Banks compared all of Bailey's specimens (no one says how many; they are stored at Virginia Polytechnic Institute, Blacksburg, today) and five more to other South Florida mainland squirrels. They decided ". . . that *matecumbei* is not sufficiently distinct to warrant recognition." But they also admit that it *is* smaller.

Here is what needs to be done:

1. Examine all the available Upper Keys specimens. There are probably not fifty in museum collections (some are at YPM and AMNH: I put them there; they were road-kills). Compare these to a good sample (another fifty) of Dade County squirrels. Separate the sexes.

2. Do the four basic measurements—total length, tail length, hind foot, and ear-from-notch—overlap so broadly that most Keys specimens cannot be separated out? Be sure to eliminate obvious juveniles, of course.

3. If most Keys specimens really are smaller, go on to the skulls. Select 20 skulls initially, five of each sex from each area. Be sure to include USNM 347444, the type-specimen of *matecumbei*. Line those skulls up

The smallest grey squirrel, *Sciurus carolinensis matecumbei*.

Joseph J. Oliver

on a dark-colored desk top and concentrate on them. Look for not just differences in size (length, width, length of the tooth row), but differences in proportions, too. Look at them in dorsal view, ventral view, and profile and you will surely find some promising possibilities.

4. Now see if the differences hold up when you go back and look at *all* the specimens. Remember, the distinction need not be absolute for sub-species: three out of four will do. Most of the differences you find will probably collapse under the added weight of more specimens, but one or two might hold up.

5. Now, write a diagnosis that will enable anyone to separate most Keys squirrels from those of the mainland. You can use a combination of characters.

Frankly, I believe proving the distinctness of *matecumbei* will be easy. I have examined few specimens indeed, and no skulls critically. I would have dismissed the color characters immediately because Hubbard and Banks said they "simply do not hold in comparison with *extimus* from southern Florida." But in the specimens at AMNH it certainly seems that the color characters hold up well. All the *extimus* I have seen had a definite ochraceous wash shading to rust on the sides of the belly. Key squirrels are somber by comparison. The Key squirrels I have seen averaged three centimeters—more than an inch—smaller than *extimus* in total length; that is about six percent. Albert Schwartz, in his 1952 thesis cited above, did report on skull characters; he believed *matecumbei* was valid.

Key squirrels are still locally abundant on Plantation Key and Key Largo, but I have not seen one on Upper Matecumbe. I would be glad to hear of a surviving population there. These squirrels do very well around human habitations, given trees and cover. They are something of a nuisance around bird feeders, but a great deal more interesting than most of the birds such feeders garner in the Keys. Like their northern relatives, they build subspherical nests of leaves as high up as they can situate them. Their diet consists of various fruits like mastic, pigeon plum, and even red mangrove. They also eat available seeds and no doubt a few insects.

A thorough study of this form would cover not just its taxonomic status and morphological characters, but its ecology and behavior as well. It is the southernmost grey squirrel and a classic example of a North American mammal that has taken up residence in a West Indian ecosystem.

References

Bailey, H. H. 1937. Change of name of *Sciurus carolinensis minutus* Bailey. *Journal of Mammalogy* 18: 516.

Hubbard, J. P., and R. C. Banks. 1970. The types and taxa of Harold H. Bailey. *Proceedings of the Biological Society of Washington* 83: 321–32.

FRUIT BAT
(Artibeus jamaicensis)

The Order Chiroptera is the second largest order of mammals. . . . Inasmuch as the group is world-wide, reaching many distant islands, and inasmuch as many species and higher groups are quite localized and some very distinctive, one might expect that a good deal of work on bat zoogeography would have been undertaken. However, such is not the case.

Karl F. Koopman, 1970

. . . the island Caribs regarded bats as their guardian spirits, and whoever killed them would fall ill.

Glover Morrill Allen, 1939

This seems to be a rather rare animal in the Keys. Maynard first recorded it in 1872, but his paper was ignored. Individuals have been found from time to time, and even brought in to the Monroe County Extension Service on Stock Island and to Florida Keys Community College. No one ever saved a specimen. Then, on 3 February 1983, construction work on the East Martello Tower roof, in Key West, flushed out some bats. One of these settled on the inside of the tower wall. Several friends of mine, including Page Brown—who specializes in bird photography—saw it there. Page photographed it, producing a color transparency subsequently accessioned at the American Museum (AMNH) in New York.

I was just a short hop away, on Middle Torch Key, stuffing road-killed bunnies and doing mark-recapture studies of rare snakes. Unfortunately, it did not occur to any of those friends that I would be interested in a bat. As this story unfolds a great lesson will be revealed: You will do a species— no matter how *seemingly* rare—much more good by collecting a voucher specimen than by soft-heartedly letting it go and coming up with just a story. That bat in the East Martello Tower was within reach of its observers. They could have had it, and in no way endangered a population by collecting that single dislodged individual. If we only had the bat, not just a photograph, what a plague of controversy would have been saved.

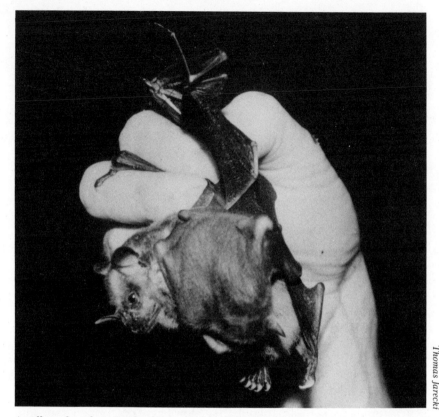

Thomas Jarecki

Antillean fruit bat, *Artibeus jamaicensis*. This female is carrying her huge infant, who appears all knees, elbows, and shoulders.

I knew what kind of bat it was as soon as I saw the slide flashed on a screen. It was the same kind of big, tailless, leaf-nosed bat Rick Warner, the Extension Service biologist at Stock Island in the 1970s, had reported to him by several people. It was an Antillean fruit bat, *Artibeus jamaicensis*. I immediately sent the slide to Dr. Karl Koopman at AMNH, the acknowledged expert on Antillean bats.

Dr. Koopman made the same identification and knew of Maynard's old Key West record and description of this as *Artibeus perspiccilalune*, which may someday prove to be the oldest, and thus correct, name for these animals. After we had published our views on Keys bats, including a black-and-white rendition of Page Brown's color slide, two Florida mammalogists, Stephen Humphrey and Larry Brown, objected. They believed our identification of the photograph was "erroneous" and that no such population can or does exist in the Keys. They believe seasonality of the Lower Keys precludes the existence of a primary figeater.

This latter point is invalid. *Artibeus* of the *jamaicensis* complex are widespread and often abundant in the Antilles and Bahamas, where they occur on islands more arid, and with more pronounced seasonality than seen in the Lower Keys. Deforestation of the hardwood hammocks has no doubt reduced an already small Lower Keys population. Continuing destruction of habitat, especially cutting the big *Ficus*—native fig trees— could surely extirpate our fruit bats. However, the big fig trees surviving right in Key West provide a far richer and more stable food source than any available—for example—on the Virgin Island of Anegada, where I have personally observed *Artibeus* populations well-known for decades. I shall return to this point.

Second, these large, tailless, leaf-nosed bats are regularly reported in the Keys and have been for years—since 1872, in fact. It is deplorable that no one has saved a specimen, but the descriptions are unmistakable. The East Martello Tower individual was not a lone, isolated occurrence.

Finally, Humphrey and Brown reject our identification on three specific points interpreted from the black-and-white print. First, they expect *Artibeus* to have light crown stripes and our bat does not. Light crown stripes may be characteristic of some (mainland, Central American?) populations, but they are not characteristic of Antillean individuals. The presence of crown stripes on a voucher from Anegada, taken many years ago, caused some biologists to suggest it represented an undescribed form. In my experience, crown stripes are present in fewer than half the Antillean individuals examined.

The remaining two Humphrey-Brown points are the crux of the case and depend wholly on interpretation of the print. What Humphrey and Brown see as a possible tail is—as they suggest—a mere fold, off-center to bat's right, in the interfemoral membrane. Last, what Humphrey and Brown call an elongate muzzle is just the shadow of the bat's face cast on the wall by the photographic flash. This latter point is easily seen in the original slide, where the brown fur of the bat's face contrasts to grey shadow on the painted brick. What seems to be a mouth line on the shadow is a small crease in the brick surface.

The photograph was taken straight on the bat's right wrist. Thus, its right forearm, pressed against the wall, shows no shadow. Its muzzle, held out from the wall and bat's left of the flash, casts a good shadow. The bat's *left* forearm, farther bat's left of the flash and also held out from the wall, casts—guess what?—an even bigger shadow. The bat seems to have a huge left arm in the print.

See, bunny huggers of the world, not even photographs are unequivocal proof. *Help the species: keep the specimen.* With a specimen in hand we might well be on our way to vindicating Maynard's form *perspiccilalune* and obtaining legal protection for a unique United States population. Without it, all we have is a spitting contest.

Since that 1983 occurrence, there have been two golden opportunities to collect specimens of Antillean fruit bats in the Keys. In 1984 a small colony—about half a dozen—moved into the attic space of the residence of Ms. Linda Pierce, a professional botanist, on Ramrod Key. An occasional individual got into the house—they made holes in the screens—and were beginning to be messy. When they left one night the Pierces sealed up the holes. No doubt about their appearance: they were big, tailless, leaf-nosed bats. By the time I got to the Pierces' though, they were gone.

Ms. Joel Beardsley on Cudjoe Key is an avid naturalist. She had often told me of seeing bats around her house, but because nighthawks are often mistaken for bats I just do not pay much attention to sight records of animals in flight. On 28 January 1986, at 7:00 p.m., I was sitting in the Beardsleys' second-floor living room when a fine, typical *Artibeus* flew against the screen, clung a second (long enough for a good look), and flew off. I was amazed. Joel and Fred Beardsley just said, "We told you so." As soon as I could coherently speak I said, "Is there a big fig tree close by?" The Beardsleys at first thought not, but then remembered a large tree a few houses east, across the road.

We were barely in sight of the tree when a fruit bat flew right down the road, directly overhead, straight to that tree. It is a huge, magnificent, native *Ficus citrifolia*. Right then, in late January, the height of the Lower Keys dry season, it was heavy with ripe, yellow fruit, which was falling and littering the ground. That should put the seasonality issue comfortably away: *Ficus* fruit all year 'round in the Keys and all the Antilles. To lack these trees an island would have to be a desert, and probably an artificial one at that.

As noted above, these are big bats, with wingspreads up to 45 centimeters. They have no tail, but rather a large, prominent nose leaf. They are our only proven representatives of the family Phyllostomidae, the New World leaf-nosed bats. Many species of this family occur in the West Indies south and east of us, and in Central and South America. Two other species seem to have occurred in the Florida Keys. One, a smaller, yellowish, short-tailed species with a long nose and no nose leaf, was photographed at Lignumvitae Key in 1984. It may have been *Phyllonycteris poeyi*, or a related species, in the opinion of Dr. Karl Koopman at AMNH. Snapshot-style photographs of it are on file there.

A small, tailless, short-faced bat without a nose leaf was captured in the Saddlebunch in 1985. By the time I heard about it, it had died and the carcass was discarded.

A free-tailed bat, *Tadarida brasiliensis,* was brought to me in 1979, captured at Bahia Honda. I gave it to Bruce Barbour, then at the Florida State Museum. The specimen never made it into the collection and no record of it exists today except in my field notes. Other kinds of bats have occasionally been reported in the Keys, but no one ever saved a specimen. The bat stories of the Keys are often much more dubious than the fish stories. At least fishermen do save an occasional specimen.

The famous bat tower established by Richter Perky, at the site that bears his name on Lower Sugarloaf, may never have been tenanted. I have been variously told that he actually brought bats over from Cuba during World War I, and that he intended to do so, but never managed it. The tower still stands, but its slats are too closely spaced to provide habitat for *Artibeus*, in my opinion.

These bats seem to pose no threat to cultivated fruits. Apparently they prefer native wild fruits of the hardwood hammocks, especially *Ficus,* the strangler fig. A lack of constantly fruiting trees in any one place no doubt keeps the population low and highly mobile. I suspect they move from key to key, roosting in big fig trees, hollow tamarinds, under palm thatch, or in buildings. Where we have studied them in the Virgin Islands, they tend to live in small groups of less than a dozen. There they produce single young, usually in the spring or summer months. They can carry their babies in flight.

The phyllostomid bats are a marvelous group of specialized animals worthy of intense study. They generally do not carry rabies (but vampires certainly do) and are not otherwise dangerous. Because raccoons do carry rabies, anyone working with live mammals in our area would be well advised to receive vaccination. Bats are rare enough in the Keys so that finding out much about them will require dedication and specialized talents. I hope someone will soon take on the job.

References

Allen, G. M. 1939. *Bats*. Cambridge, Mass.: Harvard University Press.

Baker, R. J., J. K. Jones, and D. C. Carter, eds. 1976, 1977, and 1979. *Biology of the Bats of the New World Family Phyllostomidae.* Special Publications of the Museum, Texas Tech University, Lubbock. Numbers 10, 13, and 16.

Humphrey, S. R., and L. N. Brown. 1986. Report of a new bat (Chiroptera: *Artibeus jamaicensis*) in the United States is erroneous. *Florida Scientist* 49(4): 262–63.

Koopman, K. 1970. Zoogeography of bats. In *About Bats,* B. H. Slaughter and D. W. Walton, eds. Dallas, Tex.: Southern Methodist University Press.

Lazell, J., and K. Koopman. 1985. Notes on bats of Florida's Lower Keys. *Florida Scientist* 48: 37–41.

Lazell, J., and L. Jarecki. 1985. Bats of Guana, British Virgin Islands. *American Museum Novitates* 2819: 1–7.

Maynard, C. J. 1872. Catalogue of the mammals of Florida. *Bulletin of the Essex Institute* (Massachusetts) 4: 135–40.

OPOSSUM
(DIDELPHIS V. VIRGINIANA)

*Emonge these trees is fownde that monstrous beast with a snowte
lyke a foxe, a tayle lyke a marmasette, eares lyke a batte, handes
lyke a man, and feete lyke an ape, bearing her whelpes abowte with
her in an outwarde bellye much lyke unto a great bagge or purse.*

RICHARD EDEN, 1555 (IN HARTMAN, 1952)

This is the only native marsupial in the United States. It is extremely wide-ranging, and seems to be gaining territory rapidly. At present, 'possums are abundant as far down the Keys as Lower Matecumbe. There are old records for Key Vaca, Big Pine Key, and Key West, but I have never seen a 'possum from the Middle or Lower Keys. An occasional individual may get transported along Route 1.

The 'possum has four opposable thumbs and a prehensile tail. The females have thirteen nipples inside their pouches, and often produce more young than they can feed. Baby 'possums are about the size of a honey bee when born. They crawl into their mother's pouch and attach themselves to a nipple—if they are lucky. Once furred, they often leave the pouch and cling to their mother's fur, or wrap their tails around the base of hers.

The opossum's head is a feeding structure. It is made of massive muscles that work the stout bones of the mandibles. There are 50 teeth—more than in any of our other mammals. The sensory structures—eyes, nose, tongue, ears—are present on the head, too, but the living 'possum seems not to use them for much. What little brain is contained within the skull works mostly to coordinate the hands, feet, tail, and mandibles so as to maximize feeding.

Opossums are *edificarian* wildlife: they love to live around edifices—manmade structures. They love to eat garbage, poultry, and cultivated fruits. *Homo sapiens* has made the opossum far more widespread and abundant than it could ever have been in a state of nature.

Southern 'possums are often placed in a nominal subspecies: "*D. virginiana pigra.*" The subspecies is simply not valid, as Schwartz decided back in 1952. The putative distinctions were darker coloration and a longer tail. I find the tail length of Florida specimens entirely included within the range of variation of Virginia specimens. I cannot see consistent color differences between Florida and even Massachusetts individuals, although the darkest specimens are from the South. Lots of light ones are down here, too. The pallor of Massachusetts specimens is entirely included within the range of variation of those from Florida.

To the west and south are more subspecies and additional species. Tropical America has a great assortment of opossums; three species have been recorded in Louisiana, but two were probable introductions from South or Central America.

References

Hartman, C. G. 1952. *Possums*. Austin: University of Texas Press.

Lowery, G. H. 1974. *The Mammals of Louisiana and Its Adjacent Waters*. Baton Rouge: Louisiana State University Press.

BIRDS

The different orders have, in general, no more differences between them than exist between families in other classes of vertebrates, and, anatomically generic differences are so slight that fossils are very hard to place.

ALFRED SHERWOOD ROMER, 1966

The Class Aves is distinct from all others in possessing both feathers and hollow bones: features that fossilize very well. Birds are but slightly modified reptiles, really—a twig end of the thecodont radiation that also left us crocodiles and alligators. While no living birds have teeth, quite a lot of ancient ones did. Anatomically, birds are homogeneous and quite uniform. None has lost either pair of limbs; none has lost eyes, ears, or other structures of the basic vertebrate body plan. There are no live-bearing birds, or fossorial birds, or fully aquatic birds. All birds lay eggs on land (or some structure attached to land, or floating on water), in air.

Romer's complaint about bird taxonomy is quite clearly correct. Compared to mammals (bats to whales), reptiles (turtles to snakes), or amphibians (frogs to apodans), the Class Aves lacks diversity at higher categorical levels. What ornithologists call "orders" are about what most other biologists call families. The so-called families of birds correspond roughly to genera of other animals. All living birds really belong to the same order—Neornithes—which differs from the only other real order in the class—Archeornithes—in lacking teeth.

Among other remarkable features of the Neornithes is the structure of the foot. The tarsal and metatarsal bones have either been lost completely or fused into a single remaining member, the *tibiotarsus*. Actually, birds' feet are more modified from the ancestral vertebrate plan than are their wings. The bones of the skull undergo obliterative fusion early in bird embryology, too. Although the lower jaw, for example, is truly composed of several bones (as in all vertebrates except mammals), the sutures between them are almost impossible to perceive in most adults. Birds, like mammals, have only a single systemic trunk or aorta coming from the heart, but birds have only the right (mammals only the left).

Class Aves is extremely diverse at lower categorical levels: species and subspecies. There are reckoned to be about nine thousand species of birds. Probably few are left to be discovered, because bird collectors were so avid and diligent in getting to the remote bits and pieces of the world. Nevertheless, some no doubt still lurk in the undergrowth of the tropics, and many problems are worthy of intense investigation in the Florida Keys.

Numerous popular bird guides are on the market. For the Florida Keys, I recommend the *National Geographic Society Field Guide to the Birds of North America,* written by a large committee and published by the society in 1983. Here are illustrated many Keys specialties: both quail doves, both mangrove cuckoos, both nighthawks, both noddies, and both dowitchers—but only one yellow warbler. No other guide is as good.

Second best is Roger Tory Peterson's fourth edition (1980) of *A Field Guide to the Birds.* This also illustrates most of the Keys specialties, but sometimes (quail doves) just their heads. I am quite disappointed to see the subspecies section go, especially as some of those left out are Keys favorites of mine (Gundlach's yellow warbler; Maynard's mangrove cuckoo). I agree with Peterson that many subspecies are not worth bothering with; I believe some are simply invalid. Others, however, are easily distinguished and a few may prove to be full species (I have never believed the Ipswich sparrow is a savannah sparrow). In my list below I include subspecies (the third scientific name) when I judge them to be distinguishable in the field and of major biological significance.

The Golden Press *Birds of North America* by Robbins, Bruun, and Zim—illustrated by Arthur Singer—is fine; it is preferred by some of my birding friends. It is not strong on Keys specialties and gives short shrift to subspecies, but it does include the western birds that often appear as vagrants, especially in winter, in the Keys.

The basic list I began with is that of the National Audubon Society Research Department, 115 Indian Mound Trail, Tavernier, Plantation Key, Florida 33070 (1985). I have modified and expanded it in accord with the advice of Dr. William Robertson and other colleagues (see Acknowledgments).

I do not follow the convention of capitalizing common names and I will bet you do not either in the real world (dog, horse, chicken, or human).

Annotated List of Species

All of the references cited in this list are given at the end of the Special Species accounts, which follow.

RED-THROATED LOON (*Gavia stellata*): occasional winter visitor.

COMMON LOON (*Gavia immer*): uncommon winter visitor.

PIED-BILLED GREBE (*Podilymbus podiceps*): rare winter resident.

HORNED GREBE (*Podiceps auritus*): common winter visitor; seen offshore. Very rare south of Florida Bay.

CORY'S SHEARWATER (*Calonectris diomedea*): rare transient; seen offshore.

GREATER SHEARWATER (*Puffinus gravis*): occasional transient; seen offshore.

SOOTY SHEARWATER (*Puffinus griseus*): occasional transient; seen offshore.

AUDUBON'S SHEARWATER (*Puffinus lherminieri*): uncommon summer visitor; seen offshore.

WILSON'S STORM-PETREL (*Oceanites oceanicus*): rare offshore, spring and summer.

BAND-RUMPED STORM-PETREL (*Oceanodroma castro*): two records.

WHITE-TAILED TROPICBIRD (*Phaethon lepturus*): occasional visitor; seen offshore; regular around the Dry Tortugas.

MASKED BOOBY (*Sula dactylatra*): uncommon visitor; seen offshore; regular around the Dry Tortugas, where has nested.

BROWN BOOBY (*Sula leucogaster*): uncommon permanent resident; seen offshore.

RED-FOOTED BOOBY (*Sula sula*): occasional, summer, Dry Tortugas.

NORTHERN GANNET (*Sula bassanus*): uncommon winter resident; seen offshore.

AMERICAN WHITE PELICAN (*Pelecanus erythrorhynchos*): uncommon winter resident, mostly in Florida Bay.

BROWN PELICAN (*Pelecanus occidentalis*): abundant permanent resident, but a special species: see account. Nests.

DOUBLE-CRESTED CORMORANT (*Phalacrocorax auritus*): abundant permanent resident. Nests.

ANHINGA (*Anhinga anhinga*): rare winter visitor.

MAGNIFICENT FRIGATE BIRD (*Fregata magnificens*): uncommon permanent resident. Nests in Key West National Wildlife Refuge and Tortugas. A threatened species: Robertson in Kale (1978); Halewyn and Norton (1984).

AMERICAN BITTERN (*Botaurus lentiginosus*): rare winter visitor.

76

LEAST BITTERN (*Ixobrychus exilis*): rare permanent resident. A few nesting records. See silver rice rat account *re* fresh marshes. A species of special concern: Kale (1978).

GREAT BLUE HERON (*Ardea herodias*): common permanent resident, but a special species discussed under the Special Species section. Nests.

GREAT WHITE HERON (*Ardea occidentalis*): common permanent resident. Nests. A special species worthy of a detailed account: see below.

GREAT EGRET (*Casmerodius albus*): common permanent resident. Nests. A species of special concern: Wiese in Kale (1978).

SNOWY EGRET (*Egretta thula*): common permanent resident. Nests. A species of special concern: Ogden in Kale (1978).

LITTLE BLUE HERON (*Egretta caerulea*): uncommon permanent resident. Has nested. A species of special concern: Rodgers in Kale (1978).

TRICOLORED HERON (*Egretta tricolor*): abundant permanent resident. Nests. A species of special concern: Ogden in Kale (1978).

REDDISH EGRET (*Egretta rufescens*): uncommon permanent resident. Nests. Rare, but increasing, in the state: Robertson in Kale (1978).

CATTLE EGRET (*Bubulcus ibis*): abundant permanent resident.

GREEN-BACKED HERON (*Butorides striatus*): common permanent resident. Nests.

YELLOW-CROWNED NIGHT-HERON (*Nycticorax violaceus*): common permanent resident. Nests. A species of special concern: Meyerriecks in Kale (1978).

WHITE IBIS (*Eudocimus albus*): common permanent resident. Nests. A species of special concern: Kushlan in Kale (1978).

GLOSSY IBIS (*Plegadis falcinellus*): uncommon transient. The place to see this bird is New England, at least in spring and summer. A species of special concern: Dunstan in Kale (1978).

ROSEATE SPOONBILL (*Ajaia ajaja*): uncommon permanent resident. Nests. A special species: see account below.

WOOD STORK (*Mycteria americana*): occasional winter visitor. An endangered species: Ogden in Kale (1978).

GREATER FLAMINGO (*Phoenicopterus ruber*): uncommon visitor at any time of year. Nests at Hialeah racetrack in Miami, from which site young birds have dispersed in the past. I believe it formerly nested naturally in the Keys, but many dispute this. See Robertson and Kushlan (1974).

FULVOUS WHISTLING-DUCK (*Dencrocygna bicolor*): occasional transient.

SNOW GOOSE (*Chen caerulescens*): five or six records, Boca Chica to Dry Tortugas.

CANADA GOOSE (*Branta canadensis*): two or three records.

GREEN-WINGED TEAL (*Anas crecca*): occasional winter resident.

NORTHERN PINTAIL (*Anas acuta*): occasional winter visitor.

BLUE-WINGED TEAL (*Anas discors*): common winter resident.

BAHAMA DUCK (*Anas bahamensis*): fewer than ten records.

NORTHERN SHOVELLER (*Anas clypeata*): uncommon winter resident.

GADWALL (*Anas strepera*): one Dry Tortugas record.

AMERICAN WIGEON (*Anas americana*): uncommon winter resident.

RING-NECKED DUCK (*Aythya collaris*): uncommon winter visitor.

LESSER SCAUP (*Aythya affinis*): uncommon winter visitor.

COMMON EIDER (*Somateria mollissima*): one Dry Tortugas record.

HOODED MERGANSER (*Lophodytes cucullatus*): three or four records.

RED-BREASTED MERGANSER (*Mergus serrator*): common winter resident.

BLACK VULTURE (*Coragyps atratus*): rare, five or ten records.

TURKEY VULTURE (*Cathartes aura*): common permanent resident. Nests.

OSPREY (*Pandion haliaetus*): common permanent resident. Nests—often right on phone poles along Route 1. A threatened species: Ogden in Kale (1978).

AMERICAN SWALLOW-TAILED KITE (*Elanoides forficatus*): uncommon summer resident. Nests in Upper Keys.

BALD EAGLE (*Haliaeetus leucocephalus*): rare permanent resident. Nests. A threatened species: Robertson in Kale (1978).

NORTHERN HARRIER (*Circus cyaneus*): uncommon winter resident.

SHARP-SHINNED HAWK (*Accipiter striatus*): common fall migrant and uncommon winter resident.

COOPER'S HAWK (*Accipiter cooperii*): rare transient. A species of special concern: Snyder in Kale (1978).

RED-SHOULDERED HAWK (*Buteo lineatus*): common permanent resident. Nests.

BROAD-WINGED HAWK (*Buteo platypterus*): common transient and winter resident.

SHORT-TAILED HAWK (*Buteo brachyurus*): occasional winter resident; regular on Big Pine Key, where best seen in pinelands. A rare species in the state: Ogden in Kale (1978).

SWAINSON'S HAWK (*Buteo swainsoni*): rare transient.

RED-TAILED HAWK (*Buteo jamaicensis*): occasional transient.

ROUGH-LEGGED HAWK (*Buteo lagopus*): two records.

AMERICAN KESTREL (*Falco sparverius*): common winter resident.

MERLIN (*Falco columbarius*): uncommon transient and winter resident. Listed as "status undetermined" in Florida: Wiley in Kale (1978).

PEREGRINE FALCON (*Falco peregrinus*): uncommon transient and rare winter resident. An endangered species: Snyder in Kale (1978).

BLACK RAIL (*Laterallus jamaicensis*): rare transient; three or four records.

CLAPPER RAIL (*Rallus longirostris*): common permanent resident, but a special species: see account below. Nests.

SORA (*Porzana carolina*): uncommon transient.

PURPLE GALLINULE (*Porphyrula martinica*): listed as an uncommon transient, but seems regular to me.

COMMON MOORHEN (*Gallinula chloropus*): uncommon permanent resident. Nests.

AMERICAN COOT (*Fulica americana*): common winter resident. The CARIBBEAN COOT (*F. caribea*) is a rare transient not accorded a separate entry here because my own observations in the Antilles lead me to question the validity of the species.

LIMPKIN (*Aramus guarauna*): uncommon transient. A species of special concern: Nesbitt in Kale (1978).

AMERICAN OYSTER CATCHER (*Haematopus palliatus*): rare transient; old records even to Tortugas.

BLACK-BELLIED PLOVER (*Pluvialis squatarola*): common winter resident. Some summer here: see short-billed dowitcher.

SNOWY PLOVER (*Charadrius alexandrinus*): accidental visitor. An endangered species: Woolfenden in Kale (1978).

WILSON'S PLOVER (*Charadrius wilsonia*): uncommon permanent resident. Nests.

SEMIPALMATED PLOVER (*Charadrius semipalmatus*): common winter resident.

PIPING PLOVER (*Charadrius melodus*): occasional winter resident. A species of concern: Woolfenden in Kale (1978).

KILLDEER (*Charadrius vociferus*): uncommon winter resident.

MOUNTAIN PLOVER (*Charadrius montanus*): one Key West record.

BLACK-NECKED STILT (*Himantopus mexicanus*): uncommon permanent resident. Nests.

AMERICAN AVOCET (*Recurvirostra americana*): rare winter visitor. A species of special concern: DeGange in Kale (1978).

GREATER YELLOWLEGS (*Tringa melanoleuca*): uncommon winter resident.

LESSER YELLOWLEGS (*Tringa flavipes*): common winter resident.

SOLITARY SANDPIPER (*Tringa solitaria*): occasional transient.

WILLET (*Catoptrophorus semipalmatus*): common permanent resident. Nests. Only "sandpiper" likely to be seen perched on phone wires.

SPOTTED SANDPIPER (*Actitis macularia*): listed as an uncommon permanent resident, but should not be here during the summer months. Nonmigratory resident birds could be interesting. See account of short-billed dowitcher, a special species that shares this trait.

UPLAND SANDPIPER (*Bartramia longicauda*): rare transient.

WHIMBREL (*Numenius phaeopus*): rare transient.

LONG-BILLED CURLEW (*Numenius americanus*): one Dry Tortugas record.

MARBLED GODWIT (*Limosa fedoa*): one record.

RUDDY TURNSTONE (*Arenaria interpres*): abundant permanent resident. The summering individuals are way out of their nesting range on the arctic tundra. See account of short-billed dowitcher, a special species that shares this habit.

RED KNOT (*Calidris canutus*): uncommon winter resident.

SANDERLING (*Calidris alba*): common winter resident.

SEMIPALMATED SANDPIPER (*Calidris pusilla*): uncommon permanent resident. See account of short-billed dowitcher.

WESTERN SANDPIPER (*Calidris mauri*): common winter visitor.

LEAST SANDPIPER (*Calidris minutilla*): common permanent resident. Summering birds are out of their nesting range: see account of short-billed dowitcher.

WHITE-RUMPED SANDPIPER (*Calidris fuscicollis*): uncommon transient.

BAIRD'S SANDPIPER (*Calidris bairdii*): one Dry Tortugas record.

PECTORAL SANDPIPER (*Calidris melanotus*): uncommon transient.

PURPLE SANDPIPER (*Calidris maritima*): two or three records, Key West.

DUNLIN (*Calidris alpina*): rare winter resident.

STILT SANDPIPER (*Calidris himantopus*): occasional winter resident.

BUFF-BREASTED SANDPIPER (*Tryngites subrufficollis*): a single Dry Tortugas record.

SHORT-BILLED DOWITCHER (*Limnodromus griseus*): abundant permanent resident. A special species; see account below.

LONG-BILLED DOWITCHER (*Limnodromus scolopaceus*): rare winter visitor. Good
documentation of this species in the Keys is lacking; it is more frequent
inland.

COMMON SNIPE (*Gallinago gallinago*): occasional winter resident.

WILSON'S PHALAROPE (*Phalaropus tricolor*): three or four records.

RED-NECKED PHALAROPE (*Phalaropus lobatus*): two records.

POMARINE JAEGER (*Stercorarius pomarinus*): uncommon visitor; seen offshore.

PARASITIC JAEGER (*Stercorarius parasiticus*): uncommon visitor; seen offshore.

FRANKLIN'S GULL (*Larus pipixcan*): one record, Sands Key.

LAUGHING GULL (*Larus atricilla*): abundant permanent resident. Nests.

BONAPARTE'S GULL (*Larus philadelphia*): uncommon winter resident.

RING-BILLED GULL (*Larus delawarensis*): abundant winter resident.

HERRING GULL (*Larus argentatus*): common winter resident.

ICELAND GULL (*Larus glaucoides*): three or four records.

LESSER BLACK-BACKED GULL (*Larus fuscus*): rare winter visitor.

GLAUCOUS GULL (*Larus hyperboreus*): a single Dry Tortugas record.

GREAT BLACK-BACKED GULL (*Larus marinus*): three or four records.

SABINE'S GULL (*Xema sabini*): two records, Dry Tortugas.

GULL-BILLED TERN (*Sterna nilotica*): rare transient.

CASPIAN TERN (*Sterna caspia*): uncommon winter resident. A species of special
concern: Schreiber in Kale (1978).

ROYAL TERN (*Sterna maxima*): common permanent resident. A species of special
concern: Barbour and Schreiber in Kale (1978).

SANDWICH TERN (*Sterna sandvicensis*): common winter resident. A species of
special concern: Kale (1978).

ROSEATE TERN (*Sterna dougallii*): uncommon summer resident. Nests. Declining
and possibly endangered throughout Atlantic. A threatened species: Robertson (1964); Robertson in Kale (1978); Halewyn and Norton (1984).

COMMON TERN (*Sterna hirundo*): uncommon transient and rare winter resident.

FORSTER'S TERN (*Sterna forsteri*): uncommon winter resident.

LEAST TERN (*Sterna antillarum*): common summer resident. Nests on fill spits,
road shoulders, and roofs. Still, a threatened species: Fisk in Kale (1978).

BRIDLED TERN (*Sterna anaethetus*): common summer visitor offshore on Gulf
Stream.

SOOTY TERN (*Sterna fuscata*): fairly common summer visitor offshore; nests on

Dry Tortugas. A species of special concern: Robertson (1964); Robertson in Kale (1978); Halewyn and Norton (1984).

BLACK TERN (*Chlidonias niger*): uncommon transient.

BROWN NODDY (*Anous stolidus*): uncommon summer visitor offshore; nests on Dry Tortugas. A species of special concern: Robertson (1964); Robertson in Kale (1978); Halewyn and Norton (1984).

BLACK NODDY (*Anous minutus*): rare, regular visitor to Dry Tortugas, where not yet shown to nest. A special species: see account below.

BLACK SKIMMER (*Rhynchops niger*): common winter resident. Has increased dramatically since Greene's (1946) report. A species of special concern: Barbour in Kale (1978).

DOVEKIE (*Alle alle*): four or five records.

ROCK DOVE OR STREET PIGEON (*Columba livia*): common permanent resident, especially in urbanized areas. Nests.

SCALY-NAPED PIGEON (*Columba squamosa*): two Key West records.

WHITE-CROWNED PIGEON (*Columba leucocephala*): common permanent resident; nests. A special species: see account below.

EURASIAN COLLARED DOVE (*Streptopelia decaocto*): An exotic that has colonized from released imports to the Bahamas. Rapidly increasing in numbers; already nesting throughout the Keys and recorded in Dry Tortugas. Formerly mistaken for ringed turtle-dove. See Smith (1987).

WHITE-WINGED DOVE (*Zenaida asiatica*): uncommon winter visitor. Has been established as a nesting bird on the South Florida mainland, but without human introductions would not have been likely to be more than accidental this far east.

ZENAIDA DOVE (*Zenaida aurita*): now accidental, but formerly a summer resident, nesting species. Could easily reestablish itself given the chance. Audubon (1832) gives an excellent account (in the 1978 edition cited below, vol. VI, pp. 228–32). See also Robertson in Kale (1978).

MOURNING DOVE (*Zenaida macroura*): common permanent resident. Nests.

INCA DOVE (*Columbina inca*): uncommon permanent resident in Key West, where nesting has occurred. Source of these birds may have been an introduction, or may have been a natural dispersal.

COMMON GROUND-DOVE (*Columbina passerina*): common permanent resident. Nesting records are unaccountably scarce. Greene (1946) records nesting at Stock Island. May nest on Middle Torch Key (e.g., May 1985).

KEY WEST QUAIL-DOVE (*Geotrygon chrysia*): a rare visitor today, but formerly nested in the hammocks of the Lower Keys. A special species: see account below.

RUDDY QUAIL-DOVE (*Geotrygon montana*): rare visitor, especially on Dry Tortugas. See Robertson (1980).

BLUE-HEADED QUAIL-DOVE (*Starnoenas cyanocephala*): "Accidental on the southernmost Florida Keys in summer only," Audubon, 1832 (in the 1979 edition cited below, vol. IX, pp. 217–18).

BLACK-BILLED CUCKOO (*Coccyzus erythropthalmus*): rare transient.

YELLOW-BILLED CUCKOO (*Coccyzus americanus*): uncommon summer resident. Nests.

MANGROVE CUCKOO (*Coccyzus minor*): uncommon permanent resident. Nests.

SMOOTH-BILLED ANI (*Crotophaga ani*): uncommon permanent resident. Nests.

COMMON BARN-OWL (*Tyto alba*): rare winter resident. May nest.

EASTERN SCREECH-OWL (*Otus asio*): uncommon permanent resident. Nests at least on Key Largo.

BURROWING OWL (*Athene cunicularia*): rare permanent resident. Nests at least as far down as Key Vaca. A species of special concern: Owre in Kale (1978).

BARRED OWL (*Strix varia*): rare visitor.

SHORT-EARED OWL (*Asio flammeus*): three or four records, Dry Tortugas.

COMMON NIGHTHAWK (*Chordeiles minor*): uncommon summer resident. Nests in Upper Keys.

ANTILLEAN NIGHTHAWK (*Chordeiles gundlachii*): uncommon summer resident. Nests. A special species: see account below.

LESSER NIGHTHAWK (*Chordeiles acutipennis*): rare vagrant, Dry Tortugas.

CHUCK-WILL'S-WIDOW (*Caprimulgus carolinensis*): uncommon permanent resident.

WHIP-POOR-WILL (*Caprimulgus vociferus*): occasional winter visitor.

CHIMNEY SWIFT (*Chaetura pelagica*): rare transient.

ANTILLEAN PALM SWIFT (*Tachornis phoenicobia*): rare stray, probably from Cuba (Key West).

CUBAN EMERALD (*Chlorostilbon ricordii*): rare accidental transient.

RUBY-THROATED HUMMINGBIRD (*Archilochus colubris*): uncommon winter resident.

BELTED KINGFISHER (*Ceryle alcyon*): common winter resident.

RED-BELLIED WOODPECKER (*Melanerpes carolinus*): common permanent resident. Nests.

YELLOW-BELLIED SAPSUCKER (*Sphyrapicus varius*): uncommon winter resident.

NORTHERN FLICKER (*Colaptes auratus*): uncommon permanent resident. Nests at least in Upper Keys.

PILEATED WOODPECKER (*Dryocopus pileatus*): uncommon permanent resident in Upper Keys. Has nested on Key Largo.

OLIVE-SIDED FLYCATCHER (*Contopus borealis*): one Key West record.

EASTERN WOOD-PEWEE (*Contopus virens*): uncommon transient.

EASTERN PHOEBE (*Sayornis phoebe*): occasional winter visitor.

GREAT CRESTED FLYCATCHER (*Myiarchus crinitus*): uncommon permanent resident.

LA SAGRA'S FLYCATCHER (*Myiarchus sagrae*): rare winter visitor. See Robertson and Biggs (1983).

WESTERN KINGBIRD (*Tyrannus verticalis*): uncommon winter resident.

EASTERN KINGBIRD (*Tyrannus tyrannus*): common transient.

GRAY KINGBIRD (*Tyrannus dominicensis*): common summer resident. Nests.

FORK-TAILED FLYCATCHER (*Tyrannus savana*): Sugarloaf and Tortugas records.

SCISSOR-TAILED FLYCATCHER (*Muscivora forficatus*): uncommon winter resident.

PURPLE MARTIN (*Progne subis*): uncommon transient.

TREE SWALLOW (*Tachycineta bicolor*): uncommon winter resident.

BAHAMA SWALLOW (*Tachycineta cyaneoviridis*): listed as a rare visitor, but may actually nest.

NORTHERN ROUGH-WINGED SWALLOW (*Stelgidopteryx serripennis*): uncommon transient.

BANK SWALLOW (*Riparia riparia*): uncommon transient.

CLIFF SWALLOW (*Hirundo pyrrhonota*): rare transient.

CAVE SWALLOW (*Hirundo fulva*): occasional; rather regular in Dry Tortugas.

BARN SWALLOW (*Hirundo rustica*): common permanent resident.

BLUE JAY (*Cyanocitta cristata*): occasional winter visitor: Robertson and Kushlan (1974).

AMERICAN CROW (*Corvus brachyrhynchos*): rare visitor.

FISH CROW (*Corvus ossifragus*): once an occasional resident that nested as recently as 1941 (Stock Island). See Robertson and Kushlan (1974) and Greene (1946).

CAROLINA WREN (*Thryothorus ludovicianus*): nests, north Key Largo; once, Key West.

HOUSE WREN (*Troglodytes aedon*): uncommon winter resident.

RUBY-CROWNED KINGLET (*Regulus calendula*): rare visitor.

BLUE-GRAY GNATCATCHER (*Polioptila caerulea*): common winter resident.

VEERY (*Catharus fuscescens*): occasional transient.

GRAY-CHEEKED THRUSH (*Catharus minimus*): occasional transient.

SWAINSON'S THRUSH (*Catharus ustulatus*): occasional transient.

HERMIT THRUSH (*Catharus guttatus*): rare transient.

WOOD THRUSH (*Hylocichla mustelina*): occasional transient.

AMERICAN ROBIN (*Turdus migratorius*): uncommon winter resident.

GRAY CATBIRD (*Dumetella carolinensis*): common winter resident.

NORTHERN MOCKINGBIRD (*Mimus polyglottos*): abundant permanent resident. Nests.

BAHAMA MOCKINGBIRD (*Mimus gundlachii*): occasional visitor.

BROWN THRASHER (*Toxostoma rufum*): occasional permanent resident. Nests.

WATER PIPIT (*Anthus spinoletta*): rare winter visitor.

CEDAR WAXWING (*Bombycilla cedrorum*): uncommon winter visitor.

LOGGERHEAD SHRIKE (*Lanius ludovicianus*): rare winter visitor.

EUROPEAN STARLING (*Sturnus vulgaris*): common permanent resident. An introduced, exotic nuisance detrimental to our native birds, whose nest cavities it may occupy. Nests.

WHITE-EYED VIREO (*Vireo griseus*): common permanent resident. Nests.

THICK-BILLED VIREO (*Vireo crassirostris*): one Dry Tortugas record.

SOLITARY VIREO (*Vireo solitarius*): uncommon winter resident.

YELLOW-THROATED VIREO (*Vireo flavifrons*): uncommon winter resident.

PHILADELPHIA VIREO (*Vireo philadelphicus*): rare transient.

RED-EYED VIREO (*Vireo olivaceus*): occasional transient.

BLACK-WHISKERED VIREO (*Vireo altiloquus*): abundant summer resident. Nests. Described as rare in the state: Owre in Kale (1978).

BACHMAN'S WARBLER (*Vermivora bachmanii*): may be extinct. Formerly a regular transient in migration. A special species: see account below.

BLUE-WINGED WARBLER (*Vermivora pinus*): uncommon transient. Hybrids with the next are well described in guides.

GOLDEN-WINGED WARBLER (*Vermivora chrysoptera*): uncommon transient.

TENNESSEE WARBLER (*Vermivora peregrina*): uncommon transient.

ORANGE-CROWNED WARBLER (*Vermivora celata*): uncommon winter resident.

NORTHERN PARULA (*Parula americana*): uncommon winter resident.

YELLOW WARBLER (*Dendroica petechia*): listed as an "uncommon permanent resident," but that attempts to sweep so much biology under the rug you would never walk across it. Nests. A special species: see account below.

CHESTNUT-SIDED WARBLER (*Dendroica pennsylvanica*): uncommon transient.

MAGNOLIA WARBLER (*Dendroica magnolia*): uncommon transient.

CAPE MAY WARBLER (*Dendroica tigrina*): common transient.

BLACK-THROATED BLUE WARBLER (*Dendroica caerulescens*): uncommon winter visitor.

YELLOW-RUMPED WARBLER (*Dendroica coronata*): common winter resident.

BLACK-THROATED GRAY WARBLER (*Dendroica nigrescens*): about ten records.

TOWNSEND'S WARBLER (*Dendroica townsendi*): two records, Dry Tortugas; one, Marquesas.

BLACK-THROATED GREEN WARBLER (*Dendroica virens*): uncommon winter resident.

BLACKBURNIAN WARBLER (*Dendroica fusca*): uncommon transient.

YELLOW-THROATED WARBLER (*Dendroica dominica*): uncommon winter resident. Stoddard's subspecies nests in western Florida and is regarded as rare: Stevenson in Kale (1978).

PRAIRIE WARBLER (*Dendroica discolor*): common permanent resident. Nests. A species of special concern: Stevenson in Kale (1978).

PALM WARBLER (*Dendroica palmarum*): common winter resident.

BAY-BREASTED WARBLER (*Dendroica castanea*): uncommon transient.

BLACKPOLL WARBLER (*Dendroica striata*): common spring transient. Rare in fall.

CERULEAN WARBLER (*Dendroica cerulea*): occasional transient.

BLACK-AND-WHITE WARBLER (*Mniotilla varia*): uncommon winter resident.

AMERICAN REDSTART (*Setophaga ruticilla*): common transient and uncommon winter resident.

PROTHONOTARY WARBLER (*Prothonotaria citrea*): uncommon transient.

WORM-EATING WARBLER (*Helmitheros vermivorus*): uncommon transient. A species of special concern: Stevenson in Kale (1978).

SWAINSON'S WARBLER (*Limnothlypis swainsonii*): rare transient.

OVENBIRD (*Seiurus aurocapillus*): uncommon winter resident.

NORTHERN WATERTHRUSH (*Seiurus novaboracensis*): uncommon transient and winter resident.

LOUISIANA WATERTHRUSH (*Seiurus motacilla*): rare transient. See Stevenson in Kale (1978).

KENTUCKY WARBLER (*Oporornis formosus*): uncommon transient.

CONNECTICUT WARBLER (*Oporornis agilis*): uncommon transient.

COMMON YELLOWTHROAT (*Geothlypis trichas*): common winter resident.

HOODED WARBLER (*Wilsonia citrina*): occasional transient.

WILSON'S WARBLER (*Wilsonia pusilla*): occasional winter resident.

CANADA WARBLER (*Wilsonia canadensis*): rare transient.

YELLOW-BREASTED CHAT (*Icteria virens*): rare winter visitor.

BANANAQUIT (*Coroeba flaveola*): occasional visitor.

STRIPE-HEADED TANAGER (*Spindalis zena*): occasional visitor.

SUMMER TANAGER (*Piranga rubra*): uncommon transient.

SCARLET TANAGER (*Piranga olivacea*): uncommon transient.

WESTERN TANAGER (*Piranga ludoviciana*): three or four records.

NORTHERN CARDINAL (*Cardinalis cardinalis*): common permanent resident. Nests.

ROSE-BREASTED GROSBEAK (*Pheucticus ludovicianus*): uncommon transient.

BLUE GROSBEAK (*Guiraca caerulea*): uncommon transient.

INDIGO BUNTING (*Passerina cyanea*): fairly common transient and rare winter visitor.

PAINTED BUNTING (*Passerina ciris*): uncommon winter resident.

DICKCISSEL (*Spiza americana*): occasional transient.

RUFOUS-SIDED TOWHEE (*Pipilo erythropthalmus*): rare winter visitor.

CUBAN GRASSQUIT (*Tiaris canorum*): rare accidental visitor. Illustrated in Peterson's *Field Guide* (1980).

CHIPPING SPARROW (*Spizella passerina*): three or four records.

CLAY-COLORED SPARROW (*Spizella pallida*): occasional winter visitor.

VESPER SPARROW (*Pooecetes gramineus*): three or four records.

LARK SPARROW (*Chondestes grammacus*): about ten records.

SAVANNAH SPARROW (*Passerculus sandwichensis*): uncommon winter resident.

GRASSHOPPER SPARROW (*Ammodramus savannarum*): uncommon winter resident.

SONG SPARROW (*Melospiza melodia*): two records, Dry Tortugas.

SWAMP SPARROW (*Melospiza georgiana*): occasional winter visitor.

WHITE-THROATED SPARROW (*Zonotrichia albicollis*): two records, Dry Tortugas.

DARK-EYED JUNCO (*Junco hyemalis*): three or four records, No Name and Dry Tortugas.

BOBOLINK (*Dolichonyx oryzivorus*): fairly common transient.

RED-WINGED BLACKBIRD (*Agelaius phoeniceus*): common permanent resident. Nests.

TAWNY-SHOULDERED BLACKBIRD (*Agelaius humeralis*): accidental visitor. Illustrated in Peterson's *Field Guide* (1980).

YELLOW-HEADED BLACKBIRD (*Xanthocephalus xanthocephalus*): rare winter visitor.

COMMON GRACKLE (*Quiscalus quiscula*): common summer resident. Nests.

BROWN-HEADED COWBIRD (*Molothrus ater*): visitor increasing.

GLOSSY COWBIRD (*Molothrus bonariensis*): recent colonist from Antilles; increasing and apparently breeding.

ORCHARD ORIOLE (*Icterus spurius*): fairly common transient.

NORTHERN ORIOLE (*Icterus galbula*): fairly common transient and rare winter resident. Has nested in Key West: Robertson and Kushlan (1974). Once!

PURPLE FINCH (*Carpodacus purpureus*): one Upper Keys record.

PINE SISKIN (*Carduelis pinus*): three or four records.

AMERICAN GOLDFINCH (*Carduelis tristis*): uncommon winter resident.

HOUSE SPARROW (*Passer domesticus*): uncommon permanent resident. Nests around edifices. An introduced, exotic pest.

Special Species

BROWN PELICAN
(PELECANUS OCCIDENTALIS CAROLINENSIS)

I procured specimens at different places, but nowhere as many as at Key West. There you would see them flying within pistol shot of the Wharfs, the boys frequently knocking them down with stones.

JOHN JAMES AUDUBON, 1832 (1978 EDITION 1: 130)

Today, when a pelican appears on a Florida dock, people throw fish, not rocks.

ROGER TORY PETERSON, 1983 (IN GURAVICH AND BROWN)

South Florida has been the only real stronghold for this subspecies of pelican during its catastrophic decline following the widespread application of DDT and other "hard" pesticides (the chlorinated hydrocarbons). The state bird of Louisiana, named for the Carolinas, this species was an abundant breeder along the Gulf and southern Atlantic into the 1950s. Its demise was sudden and, over much of its range, total.

When I was a boy, summering in Mississippi, hundreds or thousands of pelicans nested on the Gulf barrier islands. By 1960 they were simply gone. But pelican populations have begun to recover. Some now nest again in Louisiana and at Sand Island, Alabama. I hope in my lifetime to see the species make a total comeback. In a beautiful book, Guravich and Brown (1983) document this recovery—due, at least in part, to the efforts of one man: Ralph Heath, Jr., of the Suncoast Seabird Sanctuary (see Shachtman, 1982, and Wood and Heath, 1981).

What Heath did was make a pelican pump. He did it by rescuing maimed, mangled, and mutilated pelicans. Most biologists will tell you that trying to save Mother Nature's rejects is anti-Darwinian futility. Best to let them die. But, Heath reasoned, these birds were hardly Mother Nature's rejects. These birds were individual victims of artificial—all too human—acts of negligence or outright evil. The demise of the entire species, and so many other species, was not natural at all. It was man-made: as artificial as napalm and agent orange.

So, Heath rehabilitated the artificially devastated birds. Many flew away to resume natural lives, but many never could. For them, too badly injured to fend for themselves, life in captivity was the only life possible. But, at his tiny sanctuary at Indian Shores, Heath provided the pelicans with an environment they found acceptable for nesting. Here the one-winged, the footless, the plastic-billed, and the blind can court and squabble, nest and lay—and hatch baby pelicans.

Heath's pelican pump produces an incredible density of pelicans in a small area. The young birds have a tendency to home back to Indian Shores, but many, in keeping with the traditions of youth, disperse. Hundreds of healthy young pelicans fly away every year to repopulate their world. This is the best of all possible solutions to an endangered species problem: develop a dense population (better, a set of them) that encourages dispersal of healthy young. Then the endangered species has an excellent chance of building up its numbers and regaining lost territory while operating under all the rules and regulations of natural selection.

So, common as the brown pelican is, and always has been, in the Keys, please notice them. The story of their species is an amazing chronicle of what can happen in three decades on this planet, and of the conflicting, seemingly mutually exclusive interests of a dominant species like *Homo sapiens*. It is a modern, full-scope morality play of man and nature.

Adult brown pelicans come in two distinct color patterns: all-white head and neck, and chestnut-naped. (Do not be confused by the irrelevant all-brown juveniles.) Local folks will give you all sorts of explanations for this, but most frequently that one pattern is the male and the other the female. A quick perusal of a nesting colony will sink that theory: no matter how gay the pelicans might be, so many same-sex couples could not be producing so many chicks. So, you repair to the books.

The books will tell you that one color pattern (usually all-white) is "winter" or "breeding" plumage, and the other (usually chestnut-naped) is "summer" or "nonbreeding" plumage. Now, that does not make a whole lot of sense, since—even way down in the Virgin Islands—*most* pelican breeding takes place in the spring. If you are like me, reading things like that makes you wonder. When I wonder, I start to notice. When I notice, I jot what I see down in my field notes.

Any myopic lizard hunter can plainly see that pelicans in either and both sorts of plumage court, nest, breed, and rear young. The amount of yellow on the crown is correlated to breeding: most at the outset of courtship, least in the nonbreeding or postbreeding condition. But the all-white and chestnut-naped plumage patterns have nothing to do with that.

They have nothing to do with seasons, either. I have tallied pelicans in the Keys from November through March, in May and June, and in September. The percentages and frequencies of all-white versus chestnut-naped remain about the same. These patterns *do* seem to have to do with geography. All-white is much scarcer, or totally absent, in the British Virgin Islands, at least in March, April, June, July, and November.

Ralph Heath, Jr., at whose sanctuary you can see breeding pelicans in both patterns, says an individual bird just alternates: one moult all-white, the next chestnut-naped. It is *not*, he says, a matter of color phases, in the sense that a given bird stays in one sort of adult plumage all its life. Ralph admits you can see both or either in any month of the year.

GREAT HERONS
(GENUS ARDEA)

... we had determined to shoot nothing but the Great White Heron. ... At length ... a Heron flew right over our heads, and to make sure of it, we both fired at once. The bird came down dead. ... We now rested awhile, and breakfasted on some bisquit soaked in molasses and water, reposing under the shade of some mangroves, where the mosquitoes had a good opportunity of breaking their fast also.

JOHN JAMES AUDUBON, 1832 (1978 EDITION 3:75)

Walter Bock, in 1956, revised the heron family. Not even Bock, however, could really decide how to resolve the great white versus the great blue question, although he provides an excellent overview of the problem as then understood. Much new data have been gathered since, but we do not seem nearer a true understanding of this bizarre biological situation. Here, in bold outline, is the mare's nest:

1. Audubon described the great white heron as a distinct species, *Ardea occidentalis*. He was perfectly cognizant of color phases in herons and clearly knew, or believed he knew, that the great white was not a color phase of the great blue. He noted other anatomical differences beside plumage color, and reported (on the testimony of excellent local observers in the Keys) that the great white nested earlier than the great blue.

2. Great blue herons occur and nest throughout the range of great whites—in the Florida Keys, the Greater Antilles, and on the South American coastal islands. The great white does not have a separate geographic range. It is wholly sympatric with the great blue.

Great white heron, *Ardea occidentalis*.

Great blue heron, *Ardea herodias*.

L. Page Brown (Cornell University)

3. Great white herons and great blue herons interbreed. They produce obvious intermediates: white-head-and-necked birds called Wurdemann's herons *and*, sometimes, all white or all dark birds from a mixed pair, or an all-white bird from an all-dark pair, or vice versa.

4. Real anatomical differences distinguish *most* great whites and all great blues. There is a modal, but strongly significant, difference in nesting season: great whites nest in winter, great blues in spring. This nesting difference seems more pronounced in the Greater Antilles, where, unfortunately, both sorts of birds are so severely depleted now that any data are extremely hard to acquire.

So what is a great white heron? A species? A subspecies? Or a color morph (or "phase")? How can one find out? How does this strange situation fit into our notions about evolution, speciation, and genetics?

There is a wonderful, and undoubtedly apocryphal, story about how the AOU came to list the great white as a color phase of the great blue. I have heard this story from various sources, and because I just know it could not be true, I will tell it for fun. It seems this highly distinguished Professor in the Halls of Ivy had a graduate student full of promise and prospect. His thesis topic was comparative courting and nesting behavior of herons, and he had managed to collect most of the data, on film, that he needed for his analysis. The great white heron was the holdout. As Audubon found out from his Keys colleagues, great whites did not court and nest when the other species did, and the student had visited the Keys too late.

So, the next year he went back in time. But, utterly unlike great blues and other herons, the great whites were so behaviorally distinct they would not tolerate a blind near their nests. Try as he could, the student could not get the needed film, or even make detailed, comparable observations.

The Professor was getting anxious to see the student fly the nest. They had a serious conference. The upshot was that the great white heron was an insurmountable stumbling block. It had to go. They had to persuade the AOU to exterminate it on paper: make it a non-species.

Robertson (in Kale, 1978) provides a scholarly update of Bock's description of the situation. McHenry and Dyes (1983) provide a further, tantalizing bit of evidence solidly for the color phase interpretation. Here is my view of the biological reality.

The great white heron, *Ardea occidentalis,* is a full species, distinct from the great blue, *Ardea herodias.* It speciated in the eastern Greater Antilles in geographic isolation from great blues. Subsequently, it invaded westward, spreading into the range of Greater Antillean great blues (e.g., in Cuba), while they spread east into its range.

Hybridization between great whites and great blues took place, and still takes place, at recently colonized sites peripheral to the main range (Greater Antilles) of the great white: Florida Keys and coastal cays of South America. In these peripheral areas the great white is a newcomer (probably at Wurm glacial maximum) and selection for character divergence is as yet incomplete.

The situation is hideously complicated by the fact that there is a genuine white color morph of the great blue heron. It is *not* a great white heron, but it looks like one. There is a selective advantage in being white, or whiter, in the tropical portions of the range of the great blue that runs counter to the selective pressure for character divergence between the species. Both white morph great blues and great blue x great white hybrids enjoy some advantages. Only the straight hybrid, Wurdemann's heron, is at a reproductive disadvantage, and that is modal, not absolute.

Ms. Karin F. Zachow, when a student of Dr. Oscar Owre's at the University of Miami in 1983, prepared a master's thesis addressing the status of these birds on osteological grounds. She considered 36 characters and recognized three sorts of birds: 1) great white herons; 2) resident (southern) great blues; and 3) migratory (northern) great blues. She concluded: "I recommend a return to full species status for the Great White Heron. . . ." I believe reexamination of her material, bearing in mind my view that some all-white birds are not *Ardea occidentalis*, but are genuine white-morph *Ardea herodias*, would strengthen her conclusions.

If I am right, we need to find a way to tell real great whites from white morph great blues without skeletonizing them. We need to trace both sorts of birds, and their parents, cousins, and offspring, through life. We need to find and document nesting colonies in the Greater Antilles. There is a lot of really interesting biology going on with major implications for population biology, genetics, and evolutionary theory. Potentially superb technological spinoffs are nascent in developing sampling techniques and biochemical field procedures.

One thing is certain: the great white heron is not conceivably a sub-species of the great blue. The trinomial is inappropriate in this case. Ward's heron, *Ardea herodias wardi*, is a subspecies of the great blue. It is the form of great blue that nests in sympatry, but not necessarily in synchrony, with the great white heron in the Florida Keys.

ROSEATE SPOONBILL
(AJAIA AJAJA)

The large increase of the Florida breeding population of Roseate Spoonbills since 1940 has been attended by extension of the breeding range to the southern mainland. . . .

ROBERTSON, BREEN, AND PATTY, 1983

At one time virtually the entire breeding population of these birds in eastern North America was concentrated at a few sites in Florida Bay, especially Bottle, Stake, and Manatee Keys—all north of the Matecumbes. This is another conservation success story. Ogden (in Kale, 1978) provides a historical account; Robertson et al. (1983) report on movements of marked birds.

At three "hotspots" in the Keys these spectacular birds may often be seen. Coming down Route 1, the first site is on the right, or north, side of the highway just before the town of Layton on Long Key. Look for the ponds with weedy bottoms. The second likely prospect is on the left side of Route 1 just past the turnoff to the Big Pine Fishing Lodge. Route 1 curves right, which is north, as it comes onto Big Pine Key, so the site is on the west side of the road. Once again, look for shallow ponds. Much of this area is restored wetland reclaimed by Dr. Arthur H. Weiner from ill-advised, unnecessary, filled roads cut through the swamp. That is another conservation success story, but sadly incomplete elsewhere in the Keys as of this writing.

The third site is often the best. It is a lovely fresh-to-brackish pond at mile marker 22, left—which is south again—side of Route 1, on Cudjoe Key. The best thing to do if you want to glass this site is slow down as you approach it and pull off on the right (north) side of the highway. Cross the road and observe from the road shoulder right at the mile marker. Good birds are always here, even if not invariably spoonbills. Look closely in the cattails at the pond edge and you may see tracks of the silver rice rat, too.

Roseate spoonbill, *Ajaia ajaja*.

L. Page Brown (Cornell University)

Clapper rail, *Rallus longirostris.*

L. Page Brown (Cornell University)

CLAPPER RAIL
(RALLUS LONGIROSTRIS SUBSPECIES)

At times it is very approachable, and since little has been written regarding the species, it would be an interesting subject for additional study.

EARLE R. GREENE, 1946

Since Greene wrote his short account precious little has been added. The situation is complicated because certainly two recognizably different sorts of clapper rails may occur in the Keys, and probably three.

The endemic nesting form is *Rallus longirostris insularum.* The mainland form, on both the Gulf and Atlantic coasts, is *R. l. scotti.* North of about Palm Beach, and thus throughout the Merritt Island marshes and Cape Canaveral region, one is supposed to find *R. l. waynei.*

Scott's clapper rail occurs in the Keys as a migrant and/or late summer wanderer. Wayne's clapper rail may further complicate the picture by appearing in the Keys too. Scott's clapper rail is the "darkest of all races of this species." An update account, but leaving it as "status undetermined," is provided by Kale (1978). The mangrove clapper rail, *R. l. insularum*, is also listed as "status undetermined" by Owre (in Kale, 1978). It may come in two color phases, but is said to be greyer than *scotti* except on the breast and flanks, which are paler.

The biological relationships of these three putatively geographically replacing forms are poorly understood. One first needs to test the hypothesis that there really are three different kinds involved. It would make an excellent research topic.

SHORT-BILLED DOWITCHER
(LIMNODROMAS GRISEUS)

Although the Dowitcher is chiefly a spring and fall migrant and abundant winter resident, nonbreeding birds are seen throughout the summer....

EARLE R. GREENE, 1946

Dowitchers are of unusual interest for two disparate reasons. First, there are two species. Greene noted, "Both forms are known to occur along the Keys," but he believed they "are impossible to tell apart in the field." I bet the real bird watchers can handle the job. The presumably rare species is the western, long-billed dowitcher, *Limnodromus scolopaceus*.

Second, the short-billed *L. griseus* (at least) exemplifies a phenomenon of great potential evolutionary interest: over-summering adults far out of their breeding range. An assortment of shorebirds are apt to do this, as is noted on the preceding list. The phenomenon can lead to a remarkable sort of instant speciation. Often, out-of-range over-summering birds are juveniles. Usually over-summerers are in winter or eclipse plumage, even if adult. For these reasons, the subject of courting, mating, and nesting in our far southern climes, so remote from the arctic tundra where normal adults breed, almost never comes up.

Neoteny is the phenomenon of achieving sexual maturity while still in an otherwise juvenile condition. It is common enough in invertebrates and such vertebrates as salamanders, but rarely mentioned in the natural history of birds or mammals. Yet, I suspect it has occurred, and may readily do so again. Consider the case of the Mississippi sandhill crane, *Grus canadensis pulla*. It is currently regarded as a subspecies of the

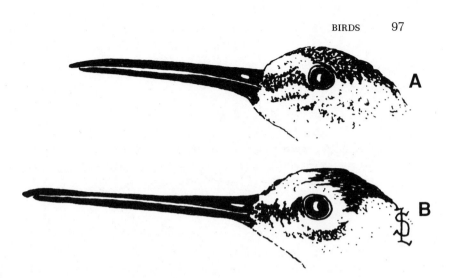

Heads of dowitchers. The short-billed has a bill *less than five times* as long as the distance from the bill base on the forehead to the center of the eye (A). It is more than five times as long in the long-billed (B)—the rarer of the two in the Keys. Drawn from actual specimens: MCZ 42357 (A) and MCZ 15468 (B).

migratory sandhill that breeds from the Great Lakes region to Saskatchewan. As the name *pulla* connotes, it is a small, somber form resembling a juvenile of the nominate, migratory sandhill. It does not migrate, but nests in the slash pine flatwoods of the Mississippi Gulf Coast. Since regular sandhills may winter there and begin courting before they fly north, opportunity for interbreeding exists. I do not know if it ever occurs, but I doubt it. Biologically, then, the Mississippi sandhill may be a full species that arose from a neotenic, nonmigratory group of regular sandhills.

Consider the mottled duck, *Anas fulvigula*, of Florida and the Gulf Coast. And the Hawaiian duck, *Anas wyvilliana*: are these not likely to be neotenic, nonmigratory mallards that achieved reproductive isolation in one quick step by breeding where they wintered? The Hawaiian duck is now threatened by introgressive hybridization with domestic and feral nonmigratory mallards brought by humans. That cheats the species of its evolutionary potential by artificially diluting the native gene pool at a crucial stage in the speciation process.

A good deal of work has been done on neoteny in insects and salamanders. Its endocrinology, biochemistry, and ethology are complex and challenging fields for investigation. So far as I know, no one has yet accepted the challenge in ornithology.

BLACK NODDY
(ANOUS MINUTUS AMERICANUS)

*The Black Noddy has a number of vocalizations, all quite different
from the Brown Noddy. . . . This species reacted to human approach
by flying overhead and calling "tick-kor-ree" and giving the staccato
rattle.*

MARY LeCROY, 1976

To date, black noddies have not been shown to nest in the Dry Tortugas,
but young birds, presumably dispersers, show up regularly there. We may
hope they will effect successful colonization in these far Keys where the
National Park Service can protect them.

Both taxonomic and conservation problems occur with black noddies.
Are they the same species as the Pacific *Anous tenuirostris*? What sorts of
real biological evidence can be brought to bear on this question? Or is it
merely a matter of opinion? More critically important, perhaps, what is
A. m. americanus? The subspecies was described in 1912 by Mathews,
who deposited his type material in the British Museum. He collected his
specimens on coastal cays off Belize (then British Honduras) where the
species has now reportedly been extirpated. In the same paper Mathews
described another form *A. m. atlanticus*, from Ascension Island.

Are these Atlantic and Caribbean forms valid subspecies? If so, which
nests at Los Roques, where Mary LeCroy studied them? Which visits the
Dry Tortugas? Does the form *americanus* survive anywhere today? If it
does, and if it visits the Dry Tortugas, can we help to save it by encouraging
black noddies to stay and nest with us? If so, how?

WHITE-CROWNED PIGEON
(COLUMBA LEUCOCEPHALA)

*They are at all times extremely shy and wary, more so in fact than
any species with which I am acquainted. The sight of a man is to
them unsupportable, perhaps on account of the continued war
waged against them, their flesh being juicy, well flavoured, and
generally tender, even in old birds.*

JOHN JAMES AUDUBON, 1832 (1978 EDITION 3: 250)

*Throughout its range the white-crowned pigeon . . . has experienced
drastic declines in numbers and has been extirpated from some
Caribbean islands.*

J. W. WILEY AND B. N. WILEY, 1979

This is a species with remarkable behavior. It is ecologically a frugivore of
the forests and hardwood hammocks, but it nests in colonies out in mangroves, over water, like so many water birds. Mass migrations depart the
Florida Keys at the end of each nesting season, seemingly to attempt
suicide in lands where they are met with a rain of shotgun pellets. Relatively few over-winter with us in relative safety. Poaching takes even some
of those.

Nevertheless, the Florida Keys support a dense population of this lovely
bird. There may now be about as many as there ever were in South Florida.
We must be genuinely zealous of our birds, and of the plight of the species
elsewhere. Dr. Oscar T. Owre (in Kale, 1978) has written eloquently on
this subject. His view is applicable to many other species we share with
our neighbors to the south:

> Reduction of season lengths and temporary suspension of hunting
> seasons where the species has been significantly reduced in numbers
> would seem to be a compelling necessity in any program of conserva-
> tion, if only to guarantee interests of future hunting. Surely the inter-
> ests of the Caribbean countries themselves in the preservation of this
> resource should be a motivating factor. *Interests of enlightened tour-
> ists* to these countries, their desire to see birds as well as their concern
> for their protection *should also be of influence to the officials formulat-
> ing the policies of Caribbean nations.*

The italics are my addition. I know the weight tourist opinion can have in
economies that depend heavily on tourism. If you travel, bear in mind
what you may be able to do.

Owre sums up, however, right back at home. I italicize his whole comment for deserved emphasis: "*In Florida the elimination or alteration of any extensive stretches of mangrove, either islets or coastal, must be subject to careful review. This is of particular importance in the Key Largo area.*"

The superb project begun by Dr. Arthur H. Weiner, using fine monies to remove abandoned roads through wild areas in the Keys, has saved uncounted white-crowns and other wildlife. These roads were conduits for poachers. Too many remain. The community of conservationists must eschew petty dissension and get on with the important tasks at hand: remove useless roads through wildlife habitat.

KEY WEST QUAIL-DOVE
(GEOTRYGON CHRYSIA)

Did ever an Egyptian pharmacopolist employ more care in embalming the most illustrious of the Pharaohs, than I did in trying to preserve from injury this most beautiful of the woodland cooers!

JOHN JAMES AUDUBON, 1832 (1978 EDITION 7: 205)

Of course Audubon had his bird in hand: he shot it so as to illustrate it accurately. A hundred and fifty years ago that hardly mattered. The Keys were lightly populated by humans and dense, remote hammocks remained aplenty. All that changed too quickly for the quail-doves. They are among the easiest birds left on Earth to shoot, and very good to eat.

I know a fair amount about quail-doves first hand. I have lived and traveled in their ranges since 1957. The closest relative of the Key West species is the bridled quail-dove (*Geotrygon mystacea*). I have assisted Dr. Robert Chipley, of The Nature Conservancy, in his detailed studies of this rare bird on Guana Island in the British Virgins—the one place we know where it is still common.

Quail-doves are much stockier than mourning or zenaida doves and much bigger than ground doves. They have a rolling, waddly gait and are reluctant to fly. They normally walk along in front of a human pursuer and hop up on objects like rocks or logs to ogle him as he approaches. We easily drive them into mist nets set at ground level, but they are infuriatingly good at bouncing out. We catch quite a few anyway.

Audubon's evidence indicates that the Key West quail-dove was rather like its tropical cousins, the white-crowned pigeon and zenaida dove: it came to the Keys in spring to nest, and most left again in the fall for the Bahamas and Greater Antilles. A few may have overwintered with us, as do some white-crowns. Today Key West quail-doves appear rarely, as obvious stragglers and dispersers, usually in the winter months. In all my travels in the Greater Antilles (including Cuba before 1960) and Bahamas, I have never encountered *Geotrygon chrysia* there in the field. I believe the species is rare to vanishing and that we should do something about it.

The Key West quail-dove should be reestablished in the Florida Keys as a breeding bird. Young birds could be netted in the Bahamas and reared in a large enclosure strategically located in a suitable Lower Keys hammock habitat. I know quail-doves can be bred in captivity: I have seen it done in the Lesser Antilles. Controlled release of young-of-the-year from the propagation enclosure to the adjacent hammock would predictably seed a wild breeding population. Artificial supplemental feeding could increase that population to high density and make a quail-dove pump—like Ralph Heath's pelican pump.

I have the site all picked out. I believe the Bahamas National Trust would cooperate with us in getting the initial stock. I am confident the plan can work. All we need is a sponsor, and this is not an expensive project. If Audubon could wax so poetic over a dead one, imagine what we would feel seeing them alive and well. Let's bring home the Key West quail-dove.

MANGROVE CUCKOO
(COCCYZUS MINOR)

The Mangrove Cuckoo is mainly of scientific interest as an example of the Antillean element in the land avifauna of Florida.

WILLIAM B. ROBERTSON (IN KALE, 1978)

This species is regarded as rare in Florida, but is fairly common in the Lower Keys. It is not an easy bird to observe, because it likes to skulk in dense hammocks and mangroves. It is a very easy bird to hear, however. Like other cuckoos in the southeast, it is thought of by many as a "rain crow." It actually seems as though its calling increases as the pressure drops and the sky darkens before a tropical downpour.

Continental mangrove cuckoo, *Coccyzus minor continentalis.*

L. Page Brown (Cornell University)

The voice is not well described in the literature. It starts out as "gaw-gaw-gaw," in less rapid succession than a yellow-bill's call, but winds down to a drawn-out "geck, geck, geck . . ." at the end. Of course, the bird is not obliged to give the entire call each time.

The native resident and nesting form of *Coccyzus minor* in South Florida is *maynardi,* named in 1887 by Ridgway, for the great naturalist C. J. Maynard, who pioneered work in Florida and first described our native fruit bat. Originally described from the Ten Thousand Islands northwest of the Everglades, this species may be expanding its range in the manner of other Antillean colonizers like the gray kingbird.

The situation is complicated by the presence of a second form, *C. m.*

continentalis. This form is supposed to be just a vagrant in South Florida. Its nesting range is from Tamaulipas, in Mexico, to Panama. Ridgway recorded it from Louisiana and Key West. Telling the two sorts of mangrove cuckoos apart is no problem. *C. m. maynardi* has an ashy throat and anterior breast (albeit shading to buff on the belly) while *C. m. continentalis* is deep ochraceous buff all over the breast.

Is *continentalis* expanding its range north and east? Will the two forms meet as breeding birds? If so, will they behave as true subspecies and interbreed freely, or as full species and maintain their distinctness?

ANTILLEAN NIGHTHAWK
(CHORDEILES GUNDLACHII)

Until June of 1941, when Roger Tory Peterson called my attention to the notes of birds over Key West, I had presumed that the migrating Night-hawks were the eastern form, minor, *and the summer resident birds, the Florida form,* chapmani. *He suggested to me the possibility of a West Indian race breeding on the keys.*

EARLE R. GREENE, 1946

It was Peterson's suggestion upon which Greene acted, more than a year later, on 11 August 1942. He finally took gun in hand, went out to Boca Chica, and shot a pair of Cuban nighthawks. In so doing, he added a new bird to the North American list and laid the foundation for the notion of *Chordeiles gundlachii* as a full species. For good measure, he took a third shot and bagged the first Key bunny to ever find its way to a museum. It would be many years, however, before the full repercussions of what Peterson pestered Earle into doing would be realized, and all of the knotty questions are far from resolved yet.

Robertson (in Kale, 1978) describes this species as rare and suggests its correct name may be *vicinus*, a name applied now to Bahamian birds. He notes: "The unresolved points are: 1) Are West Indian (and Florida Keys) nighthawks a distinct species, *Chordeiles gundlachii*, or are they subspecies of the widespread Common Nighthawk (*C. minor*) of North and Central America? 2) Are nighthawks of the Bahamas, Florida Keys and

L. Page Brown (Cornell University)

Antillean nighthawk, *Chordeiles gundlachii.*

eastern Greater Antilles subspecifically separable (as *vicinus*) from those of Cuba and Jamaica . . . ?"

He goes on: "Compounding the taxonomic uncertainty, . . . there are several characters in which West Indian birds resemble the Lesser Nighthawk (*C. acutipennis* . . .) more closely than they resemble the Common Nighthawk."

I have made a quick perusal of the situation myself and reached the following tentative conclusions: the morphological differences between the Antillean form (which nests in the Keys) and the continental form are consistent, and the vocal differences are absolute and unmistakable. *Chordeiles minor* (of Florida) occurs sympatrically with the Antillean form and the two do not interbreed despite ample opportunity to do so. Therefore the two forms are distinct species. *Chordeiles gundlachii* is the appropriate name for the Antillean nighthawk because it is older (1856) than *vicinus* (1903).

Mine is unlikely to be the last word, because as Robertson points out: "Its population dynamics as a relatively recent colonist from the West Indies and its biological relations with closely associated nighthawks of the continental form in the Upper Keys offer attractive study opportunities."

BACHMAN'S WARBLER
(VERMIVORA BACHMANII)

*On the 6th (August) I got two birds and saw about two dozen
others; broke my gun and had to leave them unmolested. On the 8th
(August) I got five and saw a dozen beside. . . . This morning
(August 9) I got two and saw another half dozen. . . . I . . . didn't
collect any of the others because I could not have taken care of the
skins.*

J. W. ATKINS, IN SCOTT (1888)

If not extinct, it is surely on the verge of extinction.

HENRY M. STEVENSON, IN KALE (1978)

In the collection of Harvard's Museum of Comparative Zoology are 73
specimens of this, the rarest of all American warblers and—if it survives at
all—one of the rarest birds on Earth. Of those, 34 are from the Florida
Keys. Of that 34, 32 were shot in Key West by J. W. Atkins between 17 July
and 25 August in the years 1888 and 1889. (The other two were shot in
March and April 1890, in the Dry Tortugas.)

It is the same story elsewhere. At the Field Museum in Chicago are 16
Atkins specimens from Key West taken between 30 July and 18 September, from 1888 to 1893. Atkins shot hundreds of Bachman's warblers at
Key West, year after year, and sold them to the museums of the world. Did
he exterminate the species? Not at all.

Graham (1985), reporting on Hal Harrison's attempts to see this gorgeous sprite of the great southern swamps, notes that egg collector Arthur
Wayne took 32 nests in South Carolina between 1906 and 1918. No nest
has been seen there since. Earle Greene apparently never saw a Bachman's warbler in the Keys between 1939 and 1942. I have never heard of
anyone seeing one here since, either.

Bachman's warbler nested in the interior from at least Missouri to
Virginia. It wintered in Cuba. It seems all the Bachman's warblers there
ever were funneled down the Florida peninsula every year, hopped the
Keys, and flew across to Cuba. The density of records is heaviest from 5 to
28 August at Key West. This is exactly the time of year all the local
naturalists and bird watchers go to Montana. It is said that not even
Peterson himself has seen *Vermivora bachmanii*. No one is looking in the
right place at the right time.

If there is even the slimmest chance of this species still living, if there is a hope of seeing this wonderful bird ever again, that hope is in the Lower Keys of Florida. Get there at the end of July and hang on into September. Search the hammocks and mangrove swamps. It will be awful. You will feed the mosquitoes breakfast, lunch, and dinner. The chiggers will handle midnight snacks. You will sweat and swelter like those of us who actually work down here. I hold out little hope, but the Florida Keys are about the only hope there is. Get a group together; get organized; gird your loins for the tortures of the damned; then go do it: look where Atkins looked, when he looked there.

CUBAN YELLOW WARBLER
(DENDROICA PETECHIA GUNDLACHII)

... credit is hereby given to Roger Tory Peterson, ... who was with me on June 15, 1941, when a male bird in full song was discovered on one of the Bay Keys. ... Peterson believed it might be a Cuban race.

EARLE R. GREENE, 1946

This is one more example of the phenomenon chronicled by Robertson and Kushlan: invasion and colonization of the continental margin by species from oceanic islands. Ecological and biogeographic theory says things are not supposed to go that way. Robertson in Kale (1978) provides a thorough review of this rare form in South Florida.

The regular eastern yellow warbler, *Dendroica petechia aestiva*, does not nest in Florida. It nests as far south as central Georgia and South Carolina. For this reason the question of species-level status and reproductive isolation of the two sorts of yellow warblers does not arise. One is likely to see both sorts together in the Keys only during the migration of the eastern form in April and September.

Song is the most reliable identification feature of the Cuban form, but striking color differences distinguish at least adult males. In eastern *D. p. aestiva* the frontal region and anterior cap are bright yellow shading to olive. In Cuban and Keys *D. p. gundlachii* the front and entire cap are duller, darker olive and the central crown has a rich chestnut tint.

As Robertson (in Kale, 1978) points out, this bird seems to be a genuine newcomer to the Keys from Cuba or the Bahamas. It sings loudly from conspicuous perches in the mangroves. J. W. Atkins (see previous account) would scarcely have missed it in the 1890s if it was here then, but he never collected it. It was not until Peterson and Greene visited the Bay Keys in 1941 that this species was noticed and added to North America's list of native breeding birds. We may hope its range will extend and expand.

Like so many other Antillean species, the Cuban yellow warbler is named for Dr. Juan Gundlach. Gundlach was born in Marburg, Germany, in 1810 and died in Cuba in 1896. He received his Ph.D. from the University of Marburg in 1838 and immediately left Europe. He reached Cuba on Christmas day, 1838, and spent most of the rest of his life there, pursuing and preserving birds, insects, mollusks—and at least sometimes virtually every other item of natural history. He was said to be well-liked by all who knew him, generous to a fault, unassuming, and diligent. He became a member of the prestigious Boston Society of Natural History in 1851, and the Naturalists Society of Boston in 1864. He was an original member of the American Ornithologists' Union in 1883.

Juan Gundlach lost his senses of taste and smell in a shooting accident as a boy. He never cared for food and ate only to satiate hunger. He never drank alcoholic beverages, or fell in love, or did anything else—apparently—except work on natural history. He published more than 40 major works and discovered and described many new species. From 1892 until his death, he was curator of the Museum of the Institute of Havana. A detailed chronicle of his life is provided by Ramsden (1915). He was probably the first and last biologist to see the Cuban macaw, *Ara tricolor,* alive.

References

Audubon, J. J. 1832 et seq. *American Wildlife Heritage* edition of 1978–9, Kent, Ohio: Volair Limited.

Bock, W. J. 1956. A generic review of the family Ardeidae (Aves). *American Museum Novitates* 1779: 1–49.

Graham, F. 1985. Celebrant of the warbler pageant. *Audubon* 87(7): 26–31.

Greene, E. R. 1942. Golden warbler nesting in Lower Florida Keys. *Auk* 59(1): 114.

————. 1943. Cuban nighthawk breeding on the Lower Florida Keys. *Auk* 60(1): 105.

————. 1946. Birds of the Lower Florida Keys. *Quarterly Journal of the Florida Academy of Sciences* 8(3): 198–265.

Guravich, D., and J. E. Brown. 1983. *Return of the Brown Pelican.* Baton Rouge: Louisiana State University Press.

Halewyn, R., and R. Norton. 1984. The status and conservation of seabirds in the Caribbean. *ICBP Technical Publication* 2: 169–222.

Hamel, P. B. 1986. *Bachman's Warbler: A Species in Peril.* Washington, D.C.: Smithsonian Institution Press.

Kale, H. W. 1978. *Rare and Endangered Biota of Florida, Volume Two, Birds.* Gainesville, Fla.: University of Florida Presses.

LeCroy, M. 1976. Bird observations in Los Roques, Venezuela. *American Museum Novitates* 2599: 1–30.

McHenry, E. N., and J. C. Dyes. 1983. First record of juvenal "white phase" great blue heron in Texas. *American Birds* 37(1): 119.

Ramsden, C. T. 1915. Juan Gundlach. *Entomological News* 26: 241–60.

Ridgway, R. 1887. *Manual of North American Birds.* Philadelphia: J. P. Lippincott.

Robertson, W. B. 1964. The terns of the Dry Tortugas. *Bulletin of the Florida State Museum* 8: 1–94.

————. 1980. Ruddy quail dove again at Dry Tortugas. *Florida Field Naturalist* 8(1): 23–24.

————, and C. W. Biggs. 1983. A West Indian *Myiarchus* in Biscayne National Park, Florida. *American Birds* 37(4): 802–4.

————, W. L. Breen, and B. Patty. 1983. Movement of marked roseate spoonbills in Florida with a review of present distribution. *Journal of Field Ornithology* 54(3): 225–36.

————, and J. Kushlan. 1974. The southern Florida avifauna. *Miami Geological Society Memoir* 2: 414–52.

Scott, W.E.D. 1888. Bachman's warbler (*Helminthophila bachmani*) at Key West, Florida. *Auk* 5: 428–30.

Shachtman, T. 1982. *The Bird Man of St. Petersburg.* New York: Macmillan.

Smith, P. W. 1987. The Eurasian collared dove arrives in the Americas. *American Birds* 41(5): 1370–79.

Sorrie, B. A. 1979. A history of the Key West Quail-Dove in the United States. *American Birds* 33(5): 728–731.

Wiley, J., and B. Wiley. 1979. The biology of the White-crowned Pigeon. *Wildlife Monographs* 64: 1–54.

Wood, F. E., and R. T. Heath, Jr. 1981. *The Suncoast Seabird Sanctuary.* St. Petersburg, Fla: Hazlett.

REPTILES AND AMPHIBIANS

Hardly a mosquito-bitten Spaniard writing home for supplies or a French sea-captain recording in his log the adventures of a shore-party but mentions "vipers" or "crocodiles," or the shocking noise the frogs made. . . . Colonizing Florida was such a strenuous matter, however, that zoölogical observation was considerably tainted with emotion, and only those forms of life which bit people, or which people could eat, elicited any enthusiasm. . . . And since the colonizing is mercifully not quite complete, current reports, too, retain enough of the old emotional taint to warm the soul of any decent herpetologist.

ARCHIE CARR, 1940

These two classes of animals are lumped together by us humans under the study of herpetology. Reptilia and Amphibia usually combine to make up a region's herpetofauna and are collectively referred to as "herps." Nevertheless, the two classes fall on either side of a line drawn through the vertebrate animals that separates profoundly distinct groups.

Reptiles—along with birds and mammals—produce a tough embryonic membrane early in life that mitigates water loss. With this membrane, the *amnion*, reptiles and their descendants were freed of spending their embryonic lives—fishlike—in water. Amphibians—along with the several classes of fishes—have no amnion and are tied to water (or saturated nests) for their embryonic lives. Most actively use gills for breathing, even if they lose them in later life.

This profound difference in embryonic structure, opening up a whole new way of life for us amniotes, makes a poor taxonomic character. It does not fossilize and is lost when the animal hatches or is born. Trying to write a general definition of Reptilia that excludes all Amphibia, or vice-versa, is

109

difficult. All we have surviving are twig-end groups left over from vast radiations of orders and families known now only from fossils. The anatomical diversity of even amphibians, when one includes fossil forms, exceeds that of either birds or mammals. The diversity of reptiles is the greatest by far for any vertebrate class, although the bony fishes (Class Osteichthys) are more speciose.

Most professional herpetologists—like me—know almost nothing about most reptiles and amphibians. The breadth and scope of herpetology staggers the imagination. Here are the best definitions for the two classes I can frame, gleaned from the references cited below (and you will need both Romer and Goin, et al., to comprehend them):

Amphibia. Anamniote vertebrates primitively possessing an otic notch in the posterior margin of the skull, and in the advanced and living forms, possessing two occipital condyles articulating the skull and vertebral column. Ten pairs of cranial nerves and mesonephric kidneys.

Reptilia. Amniote vertebrates lacking an otic notch and possessing a single occipital condyle. Lower jaw composed of several bones on each side. No feathers. Limb bones not hollow (pneumatic). Twelve pairs of cranial nerves; metanephric kidneys; no diaphragm.

Nine species of amphibians—all frogs—and at least 35 species of reptiles—representing all major groups except Rynchocephalia—inhabit the Keys. More species have been recorded and may persist as colonies of introduced or rare native forms. The major groups—orders—of reptiles are the following:

Testudinata. The turtles. Also called Chelonia. These retain the most primitive of all reptilian skull types, the anapsid: without temporal openings. However, in other respects they are perhaps the most modified of vertebrates. Their pectoral and pelvic girdles lie underneath their ribs. All but leatherbacks (Athecae) have short ribs fused to the dermal bone armor of the shell. They have a five-chambered heart quite unlike ours: it has the standard two atria but only a single ventricle; it also has two other chambers, the *conus arteriosus* and the *sinus venosus*. Males have a true mesodermal penis.

Crocodilia. The crocodiles and alligators. Also called Loricata. These huge reptiles are genuine dinosaurs, even if they look like lizards. They have diapsid skulls: two temporal openings. They have thecodont teeth: socketed in the jaw and replaced from below, like ours. They have six-chambered hearts: the basic four like ours—two atria, two ventricles—and a *conus arteriosus* and *sinus venosus*. Males have a true mesodermal penis.

Squamata. Scaly reptiles like lizards and snakes. These are derivatives of the diapsid lineage but their skulls are so various and modified that the original temporal openings can be impossible to discern. They have hearts like turtles: five chambers. The males lack a true penis, but possess paired structures called *hemipenes* unique in the animal kingdom. They are made largely of endoderm and are extensions of the hindgut or cloaca. They usually lie in the base of the tail, but for mating they are everted, one at a time, out the vent or cloacal opening. Sperm runs in a groove along each hemipenis. A squamate reptile may use one or the other alternatively, doubling his potential number of copulations over that of other animals.

The difference between lizards and snakes is vague. No snakes have well-developed limbs or digits, none have eyelids or external ears. But quite a few lizards lack some or all of those characters too, and several families of snakes, like the boas, do have hindlegs and actually use them (but not for walking).

Herps, as I call reptiles and amphibians collectively, have served as the foundation upon which most of fundamental theoretical biogeography and modern evolutionary studies have been based. This is because they are popular and eagerly sought, they do not fly around and thereby complicate the picture, they have a good fossil record, they do not usually or often thermoregulate in a way that enables them to colonize cold places, and many are brightly colored and diurnal or loudly vocal. They are fun to hunt.

The questions that remain unanswered about our Keys herps are legion. Indeed, I can tell you very little about them; most of what follows will serve to point out our appalling ignorance—despite, really, quite a large amount of work done here by a battery of fine herpetologists. The basic unanswered questions fall into three broad categories that overlap extensively:

1. *Taxonomic.* Many subspecies have been named and described in the Keys only to be subsequently sunk. A few have been resurrected, dusted off, and recognized again as valid. Authors have suggested others are probably valid; they were too hastily described, in too few words with too few specimens—but, then, perhaps too hastily sunk. More specimens and a more critical appraisal might prove them good forms. And evidence suggests that several are undescribed subspecies for which too few specimens have been available to prove the issue. Taxonomists set out to accomplish a seemingly simple, utterly elementary task: make us a list—a directory, with names and addresses—of our fellow creatures on this little planet called Earth. It is as though we published the telephone book with a randomly selected majority of the letters and digits expunged—replaced by blank space. We gave that book out to the world—industry, commerce, educators, researchers, and ordinary folks—and said: "This is *the* 'phone book, like it or not."

2. *Distributional.* These are those addresses: where the kinds of animals live. We have only foggy clues. If we do not know the names of, for example, the racers, how can we begin to know where each sort lives? Do they move?

3. *Theoretical.* How did they get where they are now? How long have they been there? What factors determine the numbers of animals on islands: island size; length of time of isolation; distance from a source of colonizers? How rapidly do species become extinct; new colonists arrive; new species evolve? Most theoretical questions ask about how life changes over time.

The situation with the herpetofauna of the Tortugas makes an interesting exemplar of our confusion and ignorance. The islands are tiny, remote, and have been a center for scientific studies for well over a century. We ought to know what lives there.

There are today seven keys in the Tortugas. Fort Jefferson, a massive brick structure, was begun on the largest of the central cluster, Garden Key, in 1847. Dr. Joseph Bassett Holder, of Lynn, Massachusetts, was posted there as surgeon about a decade later. He was a dedicated naturalist and friend of Louis Agassiz of Harvard. Dr. Holder collected specimens (and data) on the Tortugas that he faithfully pickled and crated and shipped north. His son, Charles Frederick Holder, a boy at the time, wrote of their adventures in a fictionalized account published in 1892. The geologist James E. Mills visited the Tortugas while the Holders were there. By 1862, the herpetofauna of these islands included the red-tailed skink, the five-lined skink, the racer, and the diamondback rattler. Dr. Holder shipped all but the racer; Mills got that.

Today the Tortugas support an utterly different herpetofauna. There are reef geckos, brown anoles, Indo-Pacific geckos, and Cuban tree frogs. What happened? Were the original collections real, or were the data erroneous? Did those species become extinct? Did new invaders crowd some out? We may never know.

During the first half of this century, the Carnegie Institution of Washington maintained an active research station and laboratory on Loggerhead Key, largest of the far Tortugas. From this center emanated fine papers on marine biology, a few on plants, some on the sea turtles for which the Key is named, but not one word can I find on land vertebrates other than birds.

In Dr. J. B. Holder's correspondence from Agassiz and Putnam at Harvard, I find one specific reference to the packing of reptiles, 6 August 1860. In Charles Holder's book I find only one brief aside about snakes—how hurricanes "wake" them. Admittedly, their whole focus, doctor and boys alike, was on marine organisms, which they described and depicted in detail.

What about potential habitat? All of the missing species—two skinks, the racer, and the rattler—occur regularly in buttonwood transition elsewhere in the Lower Keys. Bowman (1918) states that there was a "large stand" of buttonwood (*Conocarpus erectus*) on both Loggerhead and Garden Keys. He states the Loggerhead stand was entirely cut down for firewood, or destroyed when the Key burned over. A few representatives of the Garden Key stand persist inside the walls of Fort Jefferson, but most were cleared away for a parade ground.

Dr. William Robertson, who knows more about the Tortugas than anyone alive, doubts it. Millspaugh (1907), for example, reporting a botanical survey done in 1904, mentions no buttonwoods on Loggerhead. Of course, say I, he mentions none on Garden either, where they do occur (Bill doesn't doubt that). Millspaugh lists no *Conocarpus* at all.

Now comes Davis (1942), who shows not only buttonwoods inside the Fort, but a nice patch around a pond (number 3) over on Bush Key. None on Loggerhead, though: they haven't grown back. I will fill in some details of this mystery under the appropriate species below.

Most of our persistent problems in herpetology revolve around lack of material. People who salvage animals found dead can really help solve problems. I detail the procedure in my discussion of the rare and little-known coral snake, below.

The major herpetological references listed below are mines of information. Roger Conant's 1975 *Field Guide* alone will provide extensive details on identification, distribution, habits, and so forth, at least as things were known a decade ago. Much new information is now available, but the major points remain the same. I will assume the interested naturalist will remember these authors when reading the accounts that follow, and not repeat there citations for every species for which they are appropriate.

A final note: many of the common names I use herein are simplified versions of the "official" English names given in standard works. I agree with Wright and Wright (1949): "Normally we expect common names to come from the people at large, but with amphibians and reptiles most of the common names in literature are really bookish names." I resist the trend to formalize vernacular names and appreciate the genuine, people's names for animals they encounter. If you want to learn "real" names for different kinds of animals, learn their scientific names. Just pronounce them so anyone can spell them; it isn't hard. There is nothing legally binding or codified about English names for animals. Most inconspicuous species really have no genuine vernacular names.

References

Ashton, R. E., and P. S. Ashton. 1981. *Handbook of Reptiles and Amphibians of Florida, Part One: The Snakes.* Miami: Windward Publishing.

———. 1985. *Handbook of Reptiles and Amphibians of Florida, Part Two: Lizards, Turtles and Crocodilians.* Miami: Windward Publishing.

Bowman, H.H.M. 1918. Botanical ecology of the Dry Tortugas. *Papers of the Department of Marine Biology,* Carnegie Institution of Washington, 12:111–38.

Carr, A. 1940. A contribution to the herpetology of Florida. *University of Florida Publication, Biological Science Series* 3(1): 1–118.

————. 1952. *Handbook of Turtles.* Ithaca, N.Y.: Comstock Publishing.

————, and C. J. Goin. 1955. *Guide to the Reptiles, Amphibians, and Freshwater Fishes of Florida.* Gainesville: University of Florida Press.

Christman, S. P. 1980. Patterns of geographic variation in Florida snakes. *Bulletin of the Florida State Museum* 25(3): 157–256.

Conant, R. 1958. *A Field Guide to Reptiles and Amphibians.* Boston: Houghton Mifflin Co.

————. 1975. *A Field Guide to Reptiles and Amphibians of Eastern and Central North America.* Boston: Houghton Mifflin Co.

Davis, J. H. 1942. The ecology of the vegetation and topography of the sand keys of Florida. *Papers from the Tortugas Laboratory,* Carnegie Institution of Washington, 33: 113–95.

Duellman, W. E., and A. Schwartz. 1958. Amphibians and reptiles of southern Florida. *Bulletin of the Florida State Museum* 3: 181–324.

Goin, C., O. Goin, and G. Zug. 1978. *Introduction to Herpetology.* 3d ed. San Francisco: W. H. Freeman and Co.

Holder, C. F. 1892. *Along the Florida Reef.* New York: D. Appleton and Co.

McDiarmid, R. W., ed. 1978. *Rare and Endangered Biota of Florida, Volume Three, Reptiles and Amphibians.* Gainesville: University Presses of Florida.

Millspaugh, C. F. 1907. Flora of the sand keys of Florida. *Field Columbian Museum Publications, Botanical Series* 2(5): 191–245.

Paulson, D. R. 1968. Variation in some snakes from the Florida Keys. *Quarterly Journal of the Florida Academy of Sciences* 29(4): 295–308.

Romer, A. S. 1966. *Vertebrate Paleontology.* 3d ed. Chicago: University of Chicago Press.

Smith, H. M. 1946. *Handbook of Lizards.* Ithaca, N.Y.: Comstock Publishing.

————. 1978. *A Guide to Field Identification: Amphibians of North America.* New York: Golden Press.

————, and E. D. Brodie. 1982. *A Guide to Field Identification: Reptiles of North America.* New York: Golden Press.

Wright, A. H., and A. A. Wright. 1949. *Handbook of Frogs and Toads.* 3d ed. Ithaca, N.Y.: Comstock Publishing.

————, and ————. 1957. *Handbook of Snakes.* 2 vols. Ithaca, N.Y.: Comstock Publishing.

DIAMONDBACK RATTLESNAKE
(CROTALUS ADAMANTEUS)

An ominously impressive snake to meet in the field. . . .

ROGER CONANT, 1975

This is by orders of magnitude the largest, most common, and most widespread venomous snake in the Keys. To date, it is the only species of rattler known from the Keys. Stories of "pigmy rattlers" persist in the Lower Keys, at least, but every specimen I have ever seen was a baby diamondback.

Females bear live young in litters of up to 18, usually in summer. At birth, the babies are about 14 inches (35 cm) long. They are born loaded with venom and able to fend for themselves, although there is some indication that they remain in proximity of the mother for a short while. Diamondback rattlers can live 15 years and attain at least eight feet (224 cm), and are regularly reported to grow much larger.

The ability of these snakes to take to the open sea is near-legendary. They are frequently encountered swimming in the Keys, often headed for some distant mangrove with not a bit of dry land in it. No doubt they eat young birds that fall from their nests on the rookery groves.

Jack Watson once encountered a big fellow forty miles out from Naples when Jack was heading for the Tortugas. The total distance is about 115 miles. This snake would have been about 21 miles southwest of the nearest landfall at Cape Romano. Jack didn't say which way it was heading. He described it as floating "like a hat"—well up out of the water, coiled. I have several times observed swimming rattlers inflate their bodies to simply float passively for a while.

The giant rattlers of the Marquesas are legendary, but I have never seen a specimen taken there. Excellent naturalists like the Fords of Key West, who frequent the sand keys, have never encountered them there. Of course, Dr. Holder collected one in the Tortugas. The specimen, albeit just the severed head, is quite real. It resided at the MCZ, Harvard, for a hundred years, and was traded away to a South American museum in 1962. At that time it was simply not appreciated that there was anything remarkable about a Tortugas rattlesnake. No one has ever seen one there since.

If there really were buttonwood thickets on Loggerhead and Garden Keys, rattlers are the very sort of snake to have lived in them. With the coming of masses of people to construct Fort Jefferson, the woods could have been quickly cleared away for nothing more than cooking firewood. Surely every rattler spotted would have been slaughtered. It is not unreasonable that Dr. Holder simply got the last one. Perhaps he failed and a few still lurk in a remote baycedar thicket. I can hope.

Given the seagoing proclivities of these snakes, it is remarkable that Christman (1980) found sharp differences between Lower Keys individuals and those from the mainland. Of course, subspeciation in genetic contact does occur—not rarely, in fact. And the Lower Keys are a distinctive ecological zone.

Lower Keys rattlers have high ventral scale counts and more lower labial scales than most mainland individuals. In addition, they have less pigmentation of the upper labials and more dark pigment on the belly than is modally the case elsewhere in the range. Using a combination of characters, one might well be able to write a solid diagnosis for an undescribed, unnamed Key rattler. I am on the case, but it will take time and research. I have amassed ten specimens in MCZ over as many years (or more). There are certainly that many at FSM and other museums. Some (like that Tortugas specimen) are just heads; others are faded, flat skins I talked their owners out of. Not all can be used for all characters. I need more specimens.

I need to do an electrophoretic comparison to see if any evidence of genetic isolation is manifest. Such evidence would be mightily encouraging. It would also be useful to look at water balance and salinity tolerance. Important physiological differences can also support a taxonomic assessment (see Key mud turtle, below).

But I will not kill these magnificent creatures. Key rattlers as I know them are docile, downright gentle animals. I have rarely heard one rattle, and never known one to attack, or even defend itself by striking. I have never heard of a fatality from rattlesnake bite in the Keys. Our rattlers seem to be of a different temperament from those of the mainland.

My favorite story of rattlesnake bite in the Keys is the time one bit Jack Watson. Watson was easing up in the dead of night to make an arrest of some poachers. He was staying very low, duck-walking through the spider mangroves at the transition edge. He was watching the lights of the poachers and did not see the big rattler until he trod on its tail. It swung around and bit him in the right buttock.

Jack couldn't even curse out loud. He kept waddling low, under cover, and pulled out his well-honed knife. He stopped just long enough to reach back, blind, and cut a plug of flesh out of himself just like you or I would plug a watermelon.

He got the poachers and in the wee hours of the morning finally had time to go to the doctor's. He was pale and weak.

"Good Lord," the doctor said, "that's a terrible wound! You've lost an enormous amount of blood."

"Well," Jack said, "at least I wasn't poisoned."

In addition to the Tortugas specimen, currently in South America, there are MCZ specimens from Big Pine, Middle and Big Torch, and Summerland Keys. Duellman and Schwartz record the species also from Little Torch, Sugarloaf, Key West, and Key Largo.

Our rattlesnake deserves a detailed study. Here is a project that needs a sponsor. Keys rattlers average very large—about 180 cm in my experience. It is possible that a new form of rattlesnake—one of the largest American snakes—needs a name. I am all for patronyms; the snake and I just need a patron.

FLORIDA COTTONMOUTH
(AGKISTRODON PISCIVORUS CONANTI)

A thoroughly aroused cottonmouth throws its head upward and backward and holds its mouth wide open, revealing a white interior—origin of the name cottonmouth.

ROGER CONANT, 1975

This potentially deadly viper, named for Dr. Conant, seems scarce in the Keys. I have never encountered it in the field. Duellman and Schwartz (1958) record it from Key Vaca and Key West. I have seen specimens in private collections from Key Vaca and Grassy Key in the Middle Keys. Conant (1975) says, "apparently missing from the lower Keys despite an old record from Key West." I agree.

There are tantalizing stories about cottonmouths in the cattail marshes of the Lower Keys—one of the best from my own place: Snake Acres on Middle Torch. I have never seen one there. Carr (1940) notes cottonmouths living in abundance on small islands without fresh water, but Dunson and Freda (1985) found no evidence for unusual salt tolerance in this species. I am widely familiar with cottonmouths throughout their mainland range (and on the North Carolina Outer Banks and other barrier islands); I always associate them with fresh marshes, never salt. I suspect some insular populations are genuine evolutionary departures.

Like most other pit vipers—those with a heat sensitive facial pit like rattlesnakes—this species bears live young. Reliable figures for litters go up to fifteen. The babies are generally eight to eleven inches (up to 28 cm) at birth. They eat most any vertebrate animal, and specialize on other snakes (especially water snakes, where those occur). They may grow to over six feet (189 cm).

Baby cottonmouths are apt to be quite prettily patterned in rusty brown with a bright yellow tail tip. Like their close relatives the copperheads (which do not occur in South Florida), the cottonmouth has dark transverse bands that are *narrow in the middle of the back* and broadest at the sides. Harmless water snakes, and other harmless snakes, are just the opposite. With age, cottonmouths darken and their pattern becomes obscure. This species, so common on the mainland, is very little known and apparently genuinely rare in the Keys.

Reference

Dunson, W. A., and J. Freda. 1985. Water permeability of the skin of the amphibious snake, *Agkistrodon piscivorus*. *Journal of Herpetology* 19(1): 93–98.

EASTERN CORAL SNAKE
(MICRURUS FULVIUS)

As I was walking through a hammock . . . a large coral snake struck me savagely on the leg; when I hastily jumped away the creature lashed its body back and forth in a series of quick, lateral jerks, and in an incredibly short time had disappeared beneath the leaf-mold and sand.

ARCHIE CARR, 1940

We have heard much about its poison, but little about its place or life in nature.

WRIGHT AND WRIGHT, 1957

This is a gorgeous and genuinely deadly species: our only member of the cobra family, Elapidae. I have never encountered it in the field. Dr. Albert Schwartz collected Key Largo specimens, now at the University of Michigan. It has never been recorded on any other Key. South Florida speci-

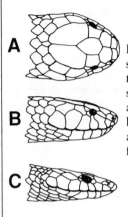

Head shape and scalation of the coral snake (A, B) and scarlet kingsnake (C). The quickest way to distinguish these brightly banded (ringed) species is by band sequence: red touches yellow in the coral snake; red and yellow are separated by black in the harmless kingsnake. But some Key Largo coral snakes may lack a yellow band. This figure is redrawn and modified from Wright and Wright (1975).

mens were designated *M. f. barbouri* by the late Karl P. Schmidt of Chicago, then the world's foremost authority on New World elapids. The subspecies was distinguished by the relative absence of black marking within the red bands. In his first (1958) edition, Conant retained this subspecies and noted that, additionally, Key Largo specimens may lack yellow bands. That same year, Duellman and Schwartz sunk the subspecies *barbouri* for inconsistency of characters.

I am not wholly convinced the form is invalid. Many South Florida coral snakes were mistakenly labeled *"barbouri,"* although clearly not that form. If the form *barbouri* is valid, it is probably restricted to the 'glades keys like Paradise Key, the type locality, and Key Largo. A combination of characters might separate *barbouri* out efficiently.

Coral snakes are easily recognized because they are the only banded or ringed snakes in our area in which red and yellow bands touch. In the harmless mimics (see scarlet kingsnake, below) the red and yellow bands are separated by black. Of course, if you meet one on Key Largo that lacks yellow bands, you will just have to know that is a possibility.

In popular imagination, coral snakes are thought of as tiny. But they are good-sized snakes, attaining nearly four feet (121 cm); it is the harmless mimics that are tiny. Of course, young coral snakes can be small too; newly hatched infants might be only seven inches (18 cm) long. They are said to lay up to a dozen eggs at about the summer solstice that hatch near the autumnal equinox.

If you encounter a brightly banded snake (bands, or rings, go *around* the body; stripes go along it) anywhere in the Keys, it is most likely to be this species. Do not mistake red ratsnakes, which have saddlelike blotches, for banded snakes. Anywhere off Key Largo, any banded snake will be a wonderful new record—if you keep it. Do not ever handle a living coral snake. They may not bite often, but once will do. Antivenin is available in Miami, but there can still be allergic complications, and time is critical. Do not take a chance.

Scoop any banded snake you find into a big jar, put a top on it, and put it in the freezer. Wait overnight; then it will be safe to handle. Let it thaw enough to bend and get it into a bottle of ordinary drugstore rubbing alcohol (70 percent isopropyl). Put a slip of cardboard with the date, locality, and your name on it—written in soft pencil—and drop it into the bottle too. Put it back in the freezer. Write me or call me; sooner or later, I will come and get it. You could become famous, in time.

You can preserve any specimen this way. Or, just keep it frozen in a tight plastic bag if it's too big for a jar. Many other biologists besides myself would come to get it. Try Florida Keys Community College, Florida International University, Key West Garden Club, or National Key Deer Wildlife Refuge. The enthusiasm you encounter will depend on who is there at the time; not all biologists love all kinds of odd animals. When you do get to someone who cares, you will feel rewarded.

Reference

Schmidt, K. P. 1928. Notes on American coral snakes. *Bulletin of the Antivenin Institute of America* 2(3): 63–64.

SCARLET KINGSNAKE
(LAMPROPELTIS TRIANGULUM ELAPSOIDES)

This small, colorful, banded serpent is a coral snake mimic.

ASHTON AND ASHTON, 1981

This species remains the most enigmatic of all our snakes. Only two specimens exist from the entire Florida Keys. One, in the collection of the University of Michigan Museum of Zoology, number 67741, is said to have been collected on Key West prior to 1929. It has 176 ventral plates, within the range of northern populations of this subspecies. Possibly for this reason many authors have ignored it. For example, Dr. Steve Christman did not include it in his consideration of variation in the species in Florida. Perhaps it does represent an erroneous locality.

However, a second specimen, at the Smithsonian, NMNH 204238, was reported on by Dr. Blair Hedges, who found it in high hammock on North Key Largo. Fitting the southern end of the cline in geographic variation exactly, it has a mere 157 ventral plates. It certainly was not introduced: the population exists.

Is it possible that our two available specimens represent two real and different populations or even different subspecies? Christman documents variation in several species in which Lower Keys individuals resemble northern Florida populations, while Upper Keys individuals fit the southern end of mainland variation. An obvious example is the racer, *Coluber constrictor*. That species is so common its variation can be reasonably analyzed (not to say it has been!).

An obvious place to look for more Lower Keys specimens, those which might support description of a new subspecies, is Big Pine Key. Stories persist of brightly banded, red, black, and yellow snakes from Big Pine. The pineland habitat is right. This is a cryptic species, scarce over most of its range, and rarely seen except on hot nights, or when flooded out by heavy rains, or when sought by dedicated herpetologists who peel the bark from rotting pine logs. Nevertheless, many diligent herpetologists, demonstrably good at finding scarlet kingsnakes, have worked Big Pine over the years, and none has ever turned up a specimen.

Archie Carr (1940) recorded scarlet kingsnakes from "mesophytic hammock," and Hedges' snake was in hardwoods. So, although most of us associate this species with pine woods, that is not necessarily the best habitat, especially in the Keys (there seem never to have been pines on Key West).

Like other kingsnakes (and coral snakes) these feed primarily on other reptiles. Skinks, especially the little brown *Scincella* (see below), and *Anolis* lizards seem to be favored items. Ashton and Ashton (1981) observed breeding in mid-April. Eggs are thought to be laid in late spring or summer and take 60–75 days to hatch. I have not seen any remarkable longevity data; most any snake this size can live a decade or more. The maximum sizes given in the literature are 27 inches, nearly 69 cm.

References

Hedges, S. B. 1977. The presence of the scarlet kingsnake, *Lampropeltis triangulum elapsoides* Holbrook (Reptilia, Serpentes, Colubridae) in the Florida Keys. *Herpetological Review* 8(4): 125–26.

Williams, K. L. 1978. Systematics and natural history of the American milk snakes, *Lampropeltis triangulum*. *Publications in Biology and Geology*, Milwaukee Public Museum, 2: 1–258.

RED RATSNAKE
(ELAPHE GUTTATA GUTTATA)

*Results of my analysis suggest that . . . these insular populations
have not differentiated substantially from the mainland
populations; thus, . . . inclusion of* E. g. rosacea *in the synonomy of*
Elaphe guttata guttata *is supported.*

JOSEPH C. MITCHELL, 1977

I agree with Mitchell's assessment, but many other excellent authorities
do not. Mitchell's paper provides in meticulous detail a description of
geographic variation in east coast red ratsnakes, or corn snakes, and the
history of the putative subspecies *"rosacea,"* called the rosy ratsnake.
Conant (1975) provides color pictures of color variants including the so-
called rosy sort, as do Ashton and Ashton (1981). Christman (1980)
intimates his feeling that the subspecies may have been unjustly sunk.
Even I admit some Keys red ratsnakes are awfully funny looking.

The term "rosy," however, is wholly misleading. These snakes are rusty,
terra cotta, or orange, but never rosy. Some are grey with yellow saddles or
side blotches. Most show less black ventral pigment and narrower or paler
borders on their dorsal saddles than do mainland specimens. In general,
these snakes are extraordinarily variable.

They are very common. Regarding them as "threatened" (Weaver in
McDiarmid, 1978) is untenable. These are snakes that abound around
human habitations: *edificarian*—of edifices—is the word. The only snake
more often encountered in the Keys is the racer, and it is rare or extirpated
in urban areas where ratsnakes are still to be found in good numbers. A
couple of dozen Keys specimens of *Elaphe guttata* are in the MCZ, for
example, no less than ten of them from Key West—type locality of *"rosacea."* This snake is of regular occurrence in Key West today, and locally
abundant there.

I have tried hard to see a way to frame a diagnosis for *"rosacea"* that
would separate it out. I am prejudiced in its favor: I would like for there to
be an endemic Keys subspecies, but I cannot see the way. Some specimens look quite different from most mainlanders, but the most classic
"rosacea" I ever saw was not the type specimen from Key West, but a
captive from Miami.

Some Key West specimens are as dark and well-patterned as some from
the extreme northern end of the range in the New Jersey Pine Barrens.

Many Lower Keys individuals show pronounced, dark, longitudinal stripes, like *Elaphe obsoleta* (see below). This was, in fact, a signal feature in the original description of *"rosacea."* Carr (1940) comments on the peculiar situation of these orange, striped individuals, seemingly intermediate between typical red and yellow ratsnakes, way down in the Lower Keys. Christman (1980) suggests plausible biological reasons for this. Many authors over the years believed *"rosacea"* and *guttata* to be distinct species, as they were originally described. This belief was supported by the occurrence of Lower Keys snakes showing both color patterns. Today we know every sort of intermediacy also occurs.

Red ratsnakes can attain very large size, to six feet (*ca* 183 cm), but I have never seen one nearly that big. A four-footer is big today. Little

Red or rosy ratsnake, *Elaphe g. guttata.* This juvenile peeks out of a cranny in oolite. Middle Torch Key.

Nancy L. Nielsen

breeding data exist for this subspecies, and none for the Keys. Wright and Wright give references to clutches up to two dozen with a July laying date. Eggs were still present in early September, so some must hatch later than that. Ashton and Ashton say they probably mate in winter or early spring and lay up to 30 eggs in late spring or early summer.

These snakes are extremely beneficial, adults feeding largely on rats and mice. The young eat treefrogs, of which we can spare plenty. They are agile climbers at all ages and love to dwell in old buildings, big trees, and walls. Because they are red or orange and boldly patterned, people are forever killing them, supposing them to be copperheads or coral snakes.

I have examined museum specimens from Key West, Sugarloaf, Summerland, Ramrod, Middle and Little Torch, Big Pine, Long Key, and Key Largo. I have color notes on a dozen individuals alive (or fresh-dead road-kills), many of which were not collected—including one from Big Torch. Duellman and Schwartz additionally record it from Bahia Honda, Boca Grande, Key Vaca, Indian Key, "13 miles north of Key West," which must really be northeast, somewhere in the Saddlebunch, Little Pine, both Upper and Lower Matecumbe, Plantation Key, and Stock Island.

Carr provides collection details on the specimen taken in 1938 in the Marquesas. That this population could have been subsequently extirpated is unlikely, as these Keys are wild and dense.

This snake is protected by law and should never be collected as a live animal. Road-kills and those killed by unknowing people should be saved to provide museum specimens.

Reference

Mitchell, J. C. 1977. Geographic variation of *Elaphe guttata* (Reptilia: Serpentes) in the Atlantic coastal plain. *Copeia* 1977(1): 33–41.

YELLOW RATSNAKE
(ELAPHE OBSOLETA QUADRIVITTATA)

But Elaphe *of the southeast will not alone be solved by a tyro borrowing all the museum material. Someone in the center of its distribution, preferably Florida, with countless live specimens can best satisfy the requirements.*

WRIGHT AND WRIGHT, 1957

Five different forms of the species *Elaphe obsoleta* have been recognized in Florida. The one from the Keys was long known as *E. o. deckerti,* named for R. F. Deckert of Jacksonville and Dade County. He wrote extensively on the herps of both areas in the teens and twenties and conferred with Wright and Wright in 1934. He did not seem to have a high opinion of the taxonomic validity of the snake named for him, and many other authors agree.

The "yellow" ratsnakes of the Keys tend to be tan to orange in ground color. They have four longitudinal stripes, like their mainland relatives, and retain their juvenile-type dorsal blotches into adulthood. Thus, they have a sort of ladder-pattern. So do a lot of yellow ratsnakes—not just young ones—from other parts of the range. I cannot frame a diagnosis that will resurrect a Keys subspecies, but Christman (1980) seemed to think it recognizable.

Ashton and Ashton regard *"deckerti"* as nothing more than a color variation. They perhaps came closest to fitting the Wrights' criteria for problem solvers, but I am not convinced they have the whole picture correctly worked out, though I accept their assessment for the Keys.

I have no reproductive data on this species from the Keys. Elsewhere they are said to mate in spring and lay up to 28 eggs in summer. The species may attain a length of seven feet (213 cm), but does not seem to ever get that large in the Keys. These snakes could certainly live 20 years. They eat rats and mice as adults, and treefrogs when young. Often they are called "chicken snakes" and they will eat young chicks and eggs. They do so much good eating rodents, however, that it seems best to forgive them a bit of poultry. This is an excellent climber.

The distribution of this form is somewhat mysterious. It is fairly common on Key Largo and occurs regularly to Lower Matecumbe. Jack Watson was just one of several astute observers who believed both yellow and red ratsnakes occurred in the Lower Keys. I picked up a shed skin on Big

Pine that seemed to have come from a well-striped, but not boldly blotched individual, and which had scales better keeled than normal in *E. guttata*. I sent it to Roger Conant, but cannot blame him if he declines to make a positive identification.

There is a magnificent *"deckerti"* sort of specimen, MCZ 31644, from Key West, collected by A. G. Elbon in 1931. Along with it came a large series of *Sphaerodactylus elegans* (elegant geckos, see below), good circumstantial evidence for the validity of the locality data. Mr. Elbon was a resident of Key West and well known to Thomas Barbour. If this species occurs below the Upper Keys, it is very rare indeed.

Reference

Brady, M. K. 1932. A new snake from Florida. *Proceedings of the Biological Society of Washington* 45: 5–8.

INDIGO SNAKE
(DRYMARCHON CORAIS COUPERI)

This gentle favorite of children and showmen is customarily thought to be an inhabitant of . . . high and dry land.

WRIGHT AND WRIGHT, 1957

This is a huge, heavy, jet black snake exceeding eight-and-a-half feet (263 cm). It is docile and usually friendly, but sometimes hisses loudly, raises its head, and flattens its neck in the vertical plane (not horizontally, like a cobra). Native Floridians are usually not afraid of indigo snakes and generally do not kill them. Indigos have a reputation for killing and eating other snakes, especially rattlers, which earns them respect and even favor. Yankees and other tourists often do not recognize this species and are prone to kill any snake they see.

The extreme popularity of indigos in the pet trade, combined with their deaths at the hands of the ignorant and on roads, has severely depleted our populations. Today only a few indigo snakes survive in the back country of the Lower Keys.

Fortunately, this species is protected by law in Florida, where designated a species of special concern (Kochman in McDiarmid, 1978). A
permit is required to keep one in captivity. Because indigos are active,
wide-ranging, diurnal snakes people are apt to encounter them. Often we
recognize a given individual we have seen around and even get to know it
a little. Indigos will eat most any other vertebrate animal, but snakes and
rats seem to be favorites. Most of our knowledge of the biology of this
species comes from captive individuals. These mate in fall or winter and
lay up to eleven eggs in May or June. The eggs take about 90 days to hatch.
Baby indigos are about 20 inches (50–60 cm) long at hatching. The babies
are irregularly marked with light grey-brown on a dark, but not black,
background. Mainland babies have rather red faces.

Keys indigos, as adults at least, seem to have little red on the face or
throat. Red persists at least on the throat of most mainland specimens. It

Indigo snake, *Drymarchon corais couperi*. This magnificent adult male, over
two meters long, is resident on Middle Torch Key.

Nancy L. Nielsen

has been informally suggested that Lower Keys indigos may be different from mainlanders in ways other than color. There are too few specimens to prove the point. Over the years, Jack Watson salvaged and pickled a dozen indigos, mostly from Big Pine Key. When Jack retired his replacement threw out all those specimens.

The only appropriate way to accumulate specimens of this species is the way Jack Watson did it: salvage road-kills. We hope no more will be killed on the roads of the Keys, but surely some will be, and they should be salvaged and preserved. MCZ has only two specimens, one each from Big Pine and Middle Torch. I have seen, examined, and photographed a Big Torch individual. Indigo snakes are reliably reported from Little Torch, Summerland, Cudjoe, Sugarloaf, and Boca Chica. Some of those, and other Keys, may be represented in museums by specimens I have not seen. Duellman and Schwartz record this species from Key Largo, as does Moler, but I have not heard it mentioned from others of the Upper or Middle Keys.

Considering the extreme popularity of the indigo snake in the pet trade, the fact that this very popularity has severely depleted populations of this species to the point of endangerment, and the resultant interest and literature generated, it is notable that the remote, isolated, possibly distinct Lower Keys populations have been wholly neglected. The references cited below reveal nothing about our Keys populations.

References

Kuntz, G. C. 1977. Endangered species: Florida indigo. *Florida Naturalist* 50(2): 15–17.
Moler, P. E. 1985. Distribution of the eastern indigo snake, *Drymarchon corais couperi*, in Florida. *Herpetological Review* 16(2): 37–38.

RACER

(COLUBER CONSTRICTOR)

Specimens from the Lower Keys have white bellies, but dorsally are just as dark as northern racers, while two from Tortugas are darker above and below than any other racers that I have ever seen.

ARCHIE CARR, 1940

The racer of the middle Florida Keys differs widely from all previously described forms.

L. NEIL BELL, 1952

These fast, agile, diurnal, fairly big, very common snakes seem to defy a rational classification. They are widely variable in color and size. Many or most are black, at least dorsally, but lots are tan to olive-brown. Three subspecies names—*priapus, paludicola,* and *haasti*—are potentially applicable to Keys populations, but for my money they may all just as well be *Coluber constrictor constrictor.*

This confusing mess sorts out as follows: southern black racers are distinguished from northern black racers (*C. c. constrictor*) by the large size of a basal hook or spine on the hemipenes of the adult male. Notwithstanding the opinions of those who believe hemipenial differences worthy of generic rank, the character is not in this case sufficient to define a subspecies. The hemipenes of most peninsular Florida males are quite strikingly different from those of New England snakes, for example, but the variation in the South is tremendous. If one would accept as a valid subspecies a form in which only most of the adult males—less than one half the population—could be identified (and I will not), then the subspecies *priapus* would be confined to peninsular Florida. The rest of the southeastern United States would be a vast zone of intergradation.

Now comes color variation in Florida. Brown racers with pale bellies occur in the Everglades, along the coast of western Florida north at least to Captiva Island, in the Upper Keys at least to Long Key, and on Merritt Island and Cape Canaveral—away to the northeast. These pale racers are called "*paludicola,*" a classic grade or polytopic ecomorph. That means they are not a single evolutionary lineage, but a convergent development adapted to a particular habitat (open marshes and sandy islands). The "*paludicola*" morph also occurs on the North Carolina Outer Banks, where it may or may not have the "*priapus*" type hemipenes.

Nancy L. Nielsen

Black racer, *Coluber constrictor.* This individual is typical of the Lower Keys populations: very dark above and below, with white on the chin and throat only. Many individuals, however, have paler venters and extensive white or grey mottling down the underside of the neck. Middle Torch Key.

Black racers occur in the Lower Keys, northeast to at least Lower Matecumbe—a paratypic locality for Bell's form *"haasti."* Thus black and brown racers overlap on at least Long and Lower Matecumbe. Bell thought *"priapus"* should have a white belly, but its describers never claimed so, and the mainland snakes I have examined (many) have slate-black bellies like northern racers. Bell claimed his *"haasti,"* centered on Big Pine Key, had a black belly, while Carr thought Lower Keys racers had white bellies.

They were both right. Young Lower Keys racers have pearly bellies that become mottled with blue-grey and further darken with age. Many adults have slate-black bellies just like the Tortugas specimen (only one from there is available to me). In 18 adults on which I have personally taken color notes, 11 had near black bellies with white confined to the chin. Three had white throats, but no white on the ventral scales proper. Four had white ventrals, ranging from the first three to the anterior 22. Two with white ventrals and one with white only on the chin had brown loreals and labials.

Facial brown is characteristic of the *"paludicola"* grade. In the Middle and Upper Keys, racers are often slate—not real satin black—above, have brown faces, and a bold, dark postocular streak. This latter feature supposedly defines a Louisiana subspecies called *"latrunculus."*

I do not understand why Neil Bell thought *"priapus"* was white-bellied, or why he thought his *"haasti"* differed from it. I do not understand why anyone ever thought *"paludicola,"* so obviously a grade of adaptation which pops up all over the map, and which partitions the *"priapus"* sorts, was a subspecies. I do not understand how the *"latrunculus"* variant got named and described. All of these putative subspecies of racers occur in the Florida Keys. What is more remarkable, they all also occur on the North Carolina Outer Banks—*all of them*. In North Carolina they are all called *Coluber constrictor constrictor.*

Racers are not constrictors. Linnaeus described the species from his desk in Sweden, and never knew the living animal. His supplier of pickled carcasses confused racers with black ratsnakes—which are constrictors—and so gave Linnaeus false information. Racers eat most any sort of vertebrate animals and are very fond of mice and rats.

In the spring racers lay up to 22 eggs that hatch in about 60 days. Baby racers, about a foot long (30 cm) at hatching, are blotched above; Keys individuals are often quite reddish brown. Ashton and Ashton give a maximum size for Florida racers as 70 inches (178 cm), but Keys specimens seldom exceed a meter in my experience.

Christman provides a lengthy discussion of variation in this species, cites all the relevant literature, and speculates on the evolution of the complexities. He does not admit to having seen as much variation in the Keys as I claim.

I have examined specimens from Key West, Tortugas, Boca Chica, Saddlebunch, Summerland, Big and Middle Torch, Big Pine, No Name, Long Key, and Key Largo. I have color notes and other data on living individuals not collected from Upper Sugarloaf. Duellman and Schwartz record the species from Upper and Lower Matecumbe, Grassy Key, and Stock Island in addition.

References

Auffenberg, W. 1955. A reconsideration of the racer, *Coluber constrictor*, in eastern United States. *Tulane Studies in Zoology* 2(6): 89–155.

Bell, L. N. 1952. A new subspecies of the racer *Coluber constrictor. Herpetologica* 8: 21.

GREEN SNAKE
(OPHEODRYS AESTIVUS CARINATUS)

There is everywhere present a beautiful green snake. It inhabits the hammocks. . . . It lies outstretched on the branches of shrubs and trees and glides along the branches from one tree to another with surprising ease.

JOHN KUNKEL SMALL, 1919 (IN WRIGHT AND WRIGHT, 1957)

This is one of the few snakes almost no Americans fear. Most recognize it as a beneficial insectivore and let it pass unmolested. This is quite correct, of course, but of some interest since in other parts of the world very long, slender, bright green snakes can be quite venomous.

We tend to think of this as a small snake because of its slender build, but it can approach four feet (116 cm) in length. It lays a half dozen eggs (up to 12 have been recorded) in summer; they seem to hatch rather quickly because hatching dates begin in July and go through September. The young are greyer than adults and about eight inches (20 cm) long. In addition to insects and spiders, these snakes have been reported to eat snails and frogs.

I do not find them common in the Lower Keys, but Carr (1940) found them "most abundant on the Upper Keys." Most of us have the impression of seasonality in this species—a summer snake, as the name *aestivus* tells. They surely do not die off annually to hatch again the next year, so they must change their habitats in the cooler, drier months and become less conspicuous.

Dr. Arnold B. Grobman has recently (1984) published his long, detailed study of geographic variation in this widespread species. He defined a new South Florida subspecies *carinatus* on the basis of more keeled dorsal scales. Specifically, check the third dorsal row at the level of the seventh ventral plate. It should always bear a strong keel in our individuals.

Grobman also noted that belly color was bright yellow in *carinatus*, paler in other populations. He believed the Keys populations were an exception, with white bellies. Carr (1940) reported that those from tropical Florida and the Keys had their bellies "entirely unpigmented." My scant color notes tell a various tale. One Middle Torch individual had white lips, chin, and venter. A second was white along mid-venter but the belly had bright yellow sides. A third, from Big Torch, had a bright butter-yellow belly.

The colors of these snakes change very rapidly after death. Their yellows fade away; their brilliant green turns to equally spectacular blue— especially in sunlight or preservative. Since our museum specimens are road-kills (we *never* kill a live one) their colors are not to be trusted if yellow seems to be lacking.

I have examined museum specimens from Key West, Summerland, Big, Little, and Middle Torch, Big Pine, Upper Matecumbe, and Key Largo. Duellman and Schwartz record it additionally from Johnston Key and Lignumvitae. Christman indicates that the scale keeling of Keys green snakes may vary with sex. Much remains to be learned about this species.

Reference

Grobman, A. B. 1984. Scutellation variation in *Opheodrys aestivus*. *Bulletin of the Florida State Museum* 29(4): 153–70.

KEY RINGNECK SNAKE
(DIADOPHIS PUNCTATUS ACRICUS)

Eight of the 14 species of snakes occurring on the Keys exhibit variation from the norm of populations on the mainland. It is clear that the isolated nature of the Keys, especially the lower ones, has served as a mechanism to promote at least a low order of speciation. . . . This high percentage appears to be unique in temperate North America.

DENNIS PAULSON, 1968

The most distinctive of the Lower Keys forms was described and named by Paulson in his 1968 paper. It is the ringless ringneck snake: *acricus* means "without a collar." Actually, some specimens do show a pretty good neck ring, but it is faded, irregular at the edges, and poorly set off ventrally from the grey head and throat. This form also has very low ventral scale counts and seems to be smaller than other subspecies of this widespread species.

In 1980 a mark-recapture procedure was established on Middle Torch Key on a two-hectare plot of old lime grove hewn from tropical hardwood hammock and buttonwood transition, now vigorously succeeding. The site, called Snake Acres, is liberally strewn with junk, including sheet metal and plywood, and I have known it as a good snake-hunting ground since 1972. On 10 January 1980, 33 pieces of cover totaling 100 m² were numbered on the tract. These have since been turned 22 times between 06:00 hrs and 11:30 hrs during January and February, 1980–1985. Eighteen Key ringneck snakes have been captured and marked. Only two were recaptured. The longest time interval between captures was 12 days; the greatest travel distance was 47 m from the original capture site. These data indicate a density point estimator of 10, with 95 percent confidence limits of slightly less than one to 36, on this tract: five per hectare. Estimates could only be made in 1980 and 1985, the only years in which recaptures occurred.

The only other snake encountered as frequently as the Key ringneck on this tract is the black racer (17 captures, one recapture). In the course of the study diamondback rattler has been recorded six times; rough green snake and red ratsnake have each been recorded twice. The following species have been recorded once: indigo snake, brown snake, and ribbon snake. The mangrove water snake also occurs on Middle Torch Key, but has not been found on Snake Acres.

All of the other snake species noted above have much larger known ranges in the Lower Keys than does *D. p. acricus*. Most are known from virtually all of the larger Keys from No Name to Key West. Several (racer, rattler, and red ratsnake) occur westward beyond Key West, often on very small Keys (Duellman and Schwartz, 1958; Christman, 1980). *Diadophis p. acricus* has not been found on No Name, Summerland, Cudjoe, or Sugarloaf keys (all mentioned as likely by Paulson, 1968) despite extensive field work there. While I do not rule out the existence of this form here or on other Keys, our failure to find it suggests that it is not widespread or common. The form may have the most restricted range of any snake in the Lower Keys.

On the Torch Keys *Diadophis p. acricus* has been found only on the edges, or in disturbed portions, of tropical hardwood hammocks. The hammock ecosystem signals the presence of fresh water because it develops only on land high enough to have a fresh water lens. Jim Hardiman first pointed out to me the close relationship of *D. p. acricus* to fresh water. Hardiman is an apiculturist and amateur entomologist who salvaged all of the Little Torch Key specimens (MCZ 156556, 156989, 158989, and 169211) at the edge of hammock L9/15. Because of the dependence of bees on fresh water, Hardiman is acutely aware of sources that remain available during the driest months, December to May.

Hammock L9/15 has extensive, permanent fresh water; hammock L9/16, on Middle Torch, has three permanent fresh water sources: one within the two hectare Snake Acres tract, and the others within a couple of

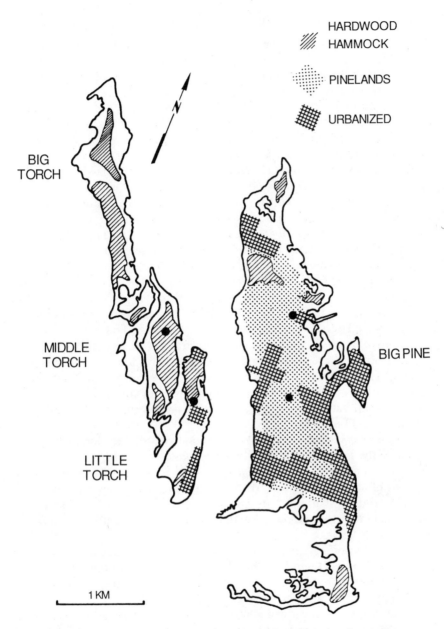

HARDWOOD
HAMMOCK

PINELANDS

URBANIZED

BIG
TORCH

MIDDLE
TORCH

LITTLE
TORCH

BIG PINE

1 KM

The presently known range of the ringless Key ringneck snake: dots indicate where specimens—salvaged as road-kills—have been collected. The principal threat to this species, and to the Key deer, for example, is increasing residential development.

hundred meters to the north and west. Thus, these two sites may be veritable centers of density for *D. p. acricus* within a very restricted overall range. Small fresh water holes are characteristic of the pineland habitats in which the Key ringneck has been found on Big Pine Key (e.g., type-locality, Paulson, 1968; Koehn Subdivision, MCZ 158988; and Key Deer Boulevard between Rt. 1 and Blue Hole, MCZ 156990).

The relatively low densities at seemingly optimal habitats and the fail-ure to recapture specimens except within a few days of initial capture may stem from several causes. Home range of individuals may be large com-pared to the two hectare study site. Fitch (1975) found that ringneck snakes moved up to 520 m and had home ranges with a long axis of about 140 m, average. Although a quarter of his snakes moved only about 10 m, travel may take many individuals out of my 100 x 200 m site. *Diadophis p. acricus* is a very small form. Discounting seemingly young snakes, appar-ently adult males averaged about 220 mm and females about 260 mm total length; these would be subadults of two or three years of age in the northern studies. In the subtropical Lower Keys, growth may well be three times as rapid as in Kansas or Michigan. Thus the Key ringneck may be characterized by a short lifespan and rapid turnover. The three individuals found in 1981 all appeared juvenile; no adults were found that year. The largest Middle Torch individual measured was a female 260 mm, total length. This would correspond to an individual entering its third winter in the northern studies, but may be much younger in the Florida Keys. A female from Little Torch, MCZ 169211, is 298 mm total length—the largest individual yet recorded.

An adult male was detained 24–48 hr on two occasions for feeding trials in captivity. He took two frogs, *Eleutherodactylus planirostris*, the first time, and a gecko, *Sphaerodactylus notatus*, plus the tails of several more geckos of the same species, the second time. Both prey species are ex-tremely abundant at the study site, with densities of up to eight per square meter of cover. It would not seem that diet could be a limiting factor for the Key ringneck snake.

Two of the other snake species present on the tract are notoriously ophiophagous: racer and indigo. Although no direct evidence of predation on ringnecks has been observed, the relative abundance of the racer is probably a major factor in ringneck ecology.

Grave problems occur in assessing factors in the demographics of small, secretive snakes. Apart from the caveats of Fitch (1975: 41–42) on mark-recapture techniques there is the question of area being sampled: Am I assaying two hectares, or merely the 100 m² of turned cover? Fitch presents evidence that, while cover eases the problems of the observer in locating snakes, it does not increase density in the habitat in Kansas. I believe increased cover might significantly improve mark-recapture procedures (if not populations) for Key ringneck snakes. It is certainly an experiment worth trying.

The scanty mark-recapture data reflect, I believe, a genuinely (if relatively) rare snake. If one assumes an average density of 5 per ha—which is possibly much too high at any distance greater than 520 m (Fitch's maximum travel distance) from a permanent fresh water source—then the approximately 1,000 ha of pinelands and hardwood hammocks left on the three Keys where *D. p. acricus* is known might support about 5,000 individuals.

Our failure to find *D. p. acricus* even on such Keys as No Name, which has extensive pinelands and the richest, wettest tropical hardwood hammocks in the Lower Keys, may well indicate a genuinely relict distribution. Evidence presented by Christman (1980) that the Lower Keys may be a refugium for ancestral character states supports the view of our endemic ringless ringneck as a relict form.

References

Blanchard, F., M. Gilreath, and F. C. Blanchard. 1979. The eastern ringneck snake (*Diadophis punctatus edwardsi*) in northern Michigan (Reptilia, Serpentes, Colubridae). *Journal of Herpetology* 13(4): 377–402.

Fitch, H. S. 1975. A demographic study of the ringneck snake (*Diadophis punctatus*) in Kansas. *Miscellaneous Publications,* University of Kansas Museum of Natural History, 62: 1–53.

Lazell, J. 1985. Geographic distribution: *Diadophis punctatus acricus. Herpetological Review* 16(4): 115.

Myers, C. W. 1965. Biology of the ringneck snake, *Diadophis punctatus* in Florida. *Bulletin of the Florida State Museum* 10(2): 43–90.

CROWNED SNAKE
(TANTILLA OOLITICA)

The fact that the type locality of the species is currently occupied by a supermarket and its attendant parking lot is symptomatic of the overall problem.

L. PORRAS AND L. D. WILSON, 1979

This delicate little creature, at home only in the deep humus of tropical hammocks, was only described and named in 1966. When Samuel R. Telford described it, he had before him only six specimens: five from Dade County and one from Key Largo. Since then at least three additional individuals have been collected on Key Largo and on Grassy Key.

All crowned snakes, genus *Tantilla*, are small and secretive. They have a distinctly set-off, enlarged, grooved fang in the rear of the mouth, but reportedly have no poison gland. In any case, they seem never to even attempt to bite humans. Instead, they please us by eating termites, spiders, and centipedes. At least, that is what our species' commoner relatives are reported to eat. No one knows for *T. oolitica*.

An excellent, beautifully illustrated account of all of Florida's crowned snakes is provided by Ashton and Ashton. Its one fault stems from the authors' habit of putting dots in the middles of counties from which species are known (rather than putting the dot where the specimen was collected). Thus, there is a big dot way over in the Everglades of western Monroe County where no member of the genus *Tantilla* has ever been found. Our *T. oolitica* does indeed occur in Monroe County, but far southeast of that dot. They show a dot on Key Largo, too.

Patterned in pastel shades of brown and pink, these snakes may attain a foot in length (29.2 cm). They are thought to lay no more than three eggs at a time, in spring. No one knows.

For some reason, this rarest of all Florida snakes, whose habitat is being or has been consumed by human development, is only listed as "threatened" (Campbell in McDiarmid, 1978). I know of no more clear-cut case of endangerment.

This state listing and classification brings up a syndrome that needs to be better understood. As Porras and Wilson point out, such listing is often counterproductive. It provides legal protection, but human predation from searching out and collecting specimens was not a problem for this species anyway. The problem is habitat destruction. Because it is virtually impossible to prove that this tiny, fossorial species occurs in a given hardwood hammock, developers need pay no attention to it. In all the arguments over North Key Largo's hammocks and their preservation, I have never heard a mention of this species.

When one was found in a Miami lot being bulldozed, it was confiscated by state game personnel and released elsewhere. Such a tiny animal has no chance, as an individual, of surviving and reproducing in a novel habitat; that animal was genetically dead as soon as its home habitat was destroyed. It should have been maintained in captivity for elucidating scientific studies. Its blood proteins should have been compared to those of its more northern relatives. A simple cell culture might have readily revealed its karyotype—number and structure of chromosomes. Karyotypic comparisons are often of great evolutionary value in determining and evaluating species-level relationships. It might have eaten in captivity, and, offered a spectrum of prey, given us totally unknown data. Had it been a female, it might well have laid eggs; female snakes store sperm and do not always need to mate prior to each clutch. A male might have lived long enough in captivity to have been provided a mate, especially if someone checked sites about to be destroyed.

By letting it go, the state authorities simply threw away knowledge for no gain at all to the endangered species.

All that *Tantilla oolitica* needs for survival is habitat. The species has no commercial value and no one except dedicated biologists who want scientific information will hunt it. Those biologists will never encounter one individual in a hundred, unless they get bulldozers and destroy habitat. Of course that is just what *no one* should be allowed to do.

This species is just one more in the lengthening list—woodrat, cotton mouse, bobcat, white-crowned pigeon, crocodile—which compel any legitimate biologist to realize that no further development should take place on Key Largo. All of the wild habitats left there—and even disturbed areas that can revert to their wild condition—should come under the Wildlife Refuge system. We have all the development we need. We are losing irreplaceable natural resources much too rapidly.

References

Porras, L., and L. D. Wilson. 1979. New distributional records for *Tantilla oolitica* Telford (Reptilia, Serpentes, Colubridae) from the Florida Keys. *Journal of Herpetology* 13(2): 218–20.

Telford, S. R. 1966. Variation among the southeastern crowned snakes, genus *Tantilla*. *Bulletin of the Florida State Museum* 10(7): 261–304.

DEKAY'S SNAKE
(Storeria dekayi victa)

The Brown Snakes from the Lower Keys are quite different from those from mainland Florida in several respects.

W. G. Weaver and S. P. Christman, 1978 (in McDiarmid)

The official common name for this form is Florida brown snake. Lots of snakes are brown. This species was named for Dr. James Ellsworth DeKay, one of America's great nineteenth-century biologists. I grew up calling it DeKay's snake, an utterly unequivocal name, and I will stick with it.

This is a little, inconspicuous species, patterned in ash-grey and slaty-blackish, especially as a punctuated middorsal pattern and neck and head markings. The scales are *keeled:* not the case with ringneck or crowned snakes. The species is reported to attain 19 inches (48 cm), but I have never seen one nearly that big. Paulson gives 39 cm for the biggest Keys specimen. A big female from Little Torch Key, MCZ 169212, salvaged as a road-kill in mid-July by Jim Hardiman, contained six embryos well-patterned and seemingly on the verge of birth. On the mainland these snakes mate in spring and summer. The females may give birth in summer, or may hold a litter over winter for spring birth.

The Lower Keys populations seem widely disjunct and genuinely scarce. Carr and Goin (1955) did not know of them. Duellman and Schwartz report it from Big Pine, No Name, and Sugarloaf. Paulson reports one specimen from Key Largo, the only record I know for the Upper to Middle Keys. MCZ has specimens from Little and Middle Torch Keys in the Lower Keys.

Middle Torch has one small area in hardwood hammock L9/16 where DeKay's snakes are encountered regularly. A few hundred meters away, where my counts of ringneck snakes have been made, this species is rare. Where I have found it there was no open fresh water.

The Lower Keys specimens have bold markings but a pallid head compared to mainlanders. Their bellies are less pigmented. They usually have two preocular scales instead of one. Other modally different scale characters are noted by Paulson (1968) and Christman (1980). Both authors felt the Lower Keys form might prove to be a valid new subspecies. There may soon be enough specimens to prove the point.

In the Lower Keys these snakes are legally protected and state-listed as threatened (Weaver and Christman, 1978, in McDiarmid), which they certainly are. Road-kills, plus analysis of the one available litter, may provide the needed data. Never pass up saving the dead body of a little brown snake.

If there is ever going to be a specific reference for Keys *Storeria dekayi*, you may have to write it. No one else has yet.

RIBBON SNAKE
(NATRIX SAURITA SACKENI)

The Lower Keys Ribbon Snake is very poorly known. Its biogeographic affinities, like several other Lower Key plants and animals, probably lie with more northern populations rather than with populations from southern Florida.

W. G. WEAVER AND S. P. CHRISTMAN, 1978 (IN MCDIARMID)

This is a slender, striped snake attaining 40 inches (101 cm). Ashton and Ashton provide an excellent color photograph. Carr and Goin (1955) did not know about the Lower Keys population. Duellman and Schwartz (1958) had it from Big Pine Key only. Paulson (1968) reported it from Cudjoe, but it is not evident that he preserved a specimen; he bemoaned the lack of color data on living specimens. Weaver and Christman (in McDiarmid, 1978) added No Name Key and state: "Most individuals have well-developed yellow or orange vertebral stripes. . . ."

I have color notes on fresh ("still twitching") road-kills, or fresh-frozen road-kills picked up by others, from Key Largo, Big Pine, Middle Torch, Cudjoe, and Upper Sugarloaf. All are now in MCZ, Harvard. Mine had somber brown vertebral stripes, hardly noticeable: just like Ashton and Ashton's picture.

Christman (1980) found a high incidence of seven supralabial scales in Lower Keys snakes, a striking departure from the mainland norm of eight. I have examined ten specimens (five are young from one female) and always get eight supralabials. Geographically, the ribbon snake is similar to DeKay's snake: a scarce disjunct, correctly listed as threatened (Weaver and Christman in McDiarmid, 1978). However, I am unable to confirm anatomical differences between Lower Keys individuals and those from the mainland.

There are fascinating features of reproductive biology, however. Ribbon snakes produce live young; usually (to the north) each embryo develops a yolk-sac sort of placental connection to the mother. The litters tend to be large: up to 20 young born in midsummer. The two known Lower Keys litters were of five and eight young each (the MCZ set from Middle Torch and a litter reported by Paulson, 1968). There was not a hint of yolk-sac placentation in the Middle Torch set. Yolk-sac placentation has been argued to be a major difference between New World *Natrix*, commonly called *"Thamnophis," "Regina,"* or *"Nerodia,"* and some of those in the Old World, like *Natrix natrix*. I will have lots more to say about this classification under the following species.

I have marked this species in my Snake Acres census counts (see ringneck snake, above), but never recaptured a specimen. The live specimen examined and released did not differ in coloration from fresh roadkills. This species seems tightly tied to open fresh water habitats: the rare, tiny marshes left in the Lower Keys.

The situation remarked by Weaver and Christman and documented extensively by Christman, of Lower Keys populations looking like northern—not southern—mainlanders is striking. Not just this species, but racers, rattlesnakes, mud turtles, and skinks show this pattern. Christman suggests that the Lower Keys were much higher in the recent past and have eroded rapidly (see The Land, above). If this were true, relictual populations might persist here and on the highland Sangamonian islands of central and north Florida. It is a most intriguing notion, but needs far more detailed knowledge of geology than I can bring to bear. Conservatively, I am most skeptical.

Reference

Moler, P. E. 1979. Geographic distribution: *Thamnophis sauritus sackeni* (Peninsula Ribbon Snake). *Herpetological Review* 10(3): 103.

MANGROVE SNAKE
(NATRIX COMPRESSICAUDA)

Natrix s. compressicauda *is apparently the only North American reptile or amphibian that has established itself in the West Indies; it has been found on the north-central coast of Cuba.*

ARCHIE CARR, 1940

This species is widespread in mangrove swamps along the north coast of Cuba, throughout the Keys, and at least as far north up the Florida Peninsula as mangroves persist. It is common and—Oh, miraculous!— pretty well studied. It is a fascinating animal in details of its biology, but before getting to that I need to dispense with the nagging problem of the generic name *Natrix*, which most of my colleagues no longer use for any American snakes.

In America today it has become customary to recognize three different genera of snakes where one would do. The three are *"Nerodia"* for our water snakes (and mangrove and saltmarsh snakes), *"Regina"* for some slender, striped species that eat crayfishes (which lots of the others do too), and *"Thamnophis"* for garter and ribbon snakes. These groups differ from each other only modally—on average—not absolutely. For this reason alone they are not valid genera.

Robin Lawson, at Louisiana State University, has recently published an analysis of the relationships of some American natricine snakes based on some selected aspects of molecular biology. The results destroy both *"Regina"* and *"Thamnophis"* as genera, although it is suggested that the latter might stand as a subgenus. Lawson suggests that only a few *"Thamnophis"* have divided anal plates; however, garter and ribbon snakes often have divided anal plates (see Lazell, 1976). Lawson and most American snake doctors are unfamiliar with Eurasian—especially Chinese—species. They have not had the opportunities to learn about these snakes in situ, or even in the world's largest reptile collection at MCZ, Harvard.

Natrix natrix, the European grass snake, is the type species of its genus. *Natrix natrix* is a garter snake. Anyone passingly familiar with, for example, the *"pallidula"* morph of the common eastern garter snake, or the western garter snakes like *elegans* or *couchi,* who then got to know *Natrix natrix* would not doubt that. Nevertheless, many scientists classify *Natrix* incorrectly. Edmund Malnate, for example, stated of the type species: "The . . . terrestrial adaptations of this species make it a poor choice . . . to represent this genus of aquatic forms." Malnate was motivated by the untenable belief that the genus *Natrix* was somehow too big—"unwieldy" was his word. There is no proper size for a genus. Genera—and families and orders—can contain one or thousands of species.

Anyway, the only difference was claimed to be that American garter snakes have single anal plates and members of *Natrix* have divided ones. The anal plate character is simply not true. In some populations of eastern garter snakes 60 percent of individuals have divided anal plates. The reverse is true too: many individuals of species that usually have divided anal plates have single ones. The condition of the anal plate cannot be used to define groups at any level.

In Eurasia and Australasia there are many species of *Natrix*. Some lay eggs, some bear live young. In his attempt to improve natricine taxonomy Malnate split the Old World forms into many spurious genera. The characters he used were inconsistent and he knew it. He stated over and over again within that paper that species after species was an "exception" to his stated generic diagnoses. The paper was uncritically accepted by most American herpetologists.

Based on the acceptance of that split, Rossman and Eberle (1977) further fragmented *Natrix,* putting the American water snakes in "*Nerodia*" on the basis of a karyotypic (chromosome) difference. They knew the karyotypes of only a few species. Their claim that hemipenial differences support this split is untrue, as revealed by close reading of Malnate's paper and examination of specimens.

It is all a house of cards. The entire complex of "*Thamnophis,*" "*Regina,*" and "*Nerodia,*" as well as Malnate's plethora of Old World genera, is simply one group: *Natrix*. That is all that can be defined or rationally defended. I have further discussed requirements for valid genera under The Fauna, above.

Our mangrove snake is a classic representative of the genus *Natrix*. It is more lightly built and slender than some of the big species, but not so gracile as the slender extreme of the genus, represented by our ribbon snake. It may be longitudinally striped or transversely banded, or both. Our mangrove snake is far more aquatic than a European grass snake (which resembles our ribbon snake in habits), and may set the outer limit for the genus in marine adaptation. Nevertheless, it lacks a salt gland; it cannot drink seawater and survive. Although Dunson (1980) first showed the physiological differences in skin that help mangrove snakes survive in a saline environment, it was shown by Miller (1985) that they drink rain water (or even potentially dew) and avoid salt intake behaviorally.

Mangrove snakes climb well. They are apt to suspend themselves in low shrubs like mangroves with the head angled down. Rain water trickles down the snake's body and accumulates at the slightly parted lips. This water the snake drinks. It will not drink seawater. It would *rather* drink falling water drops than drink fresh water out of a bowl.

During the Florida Keys dry season—"winter"—there can be soaking dews. Mangrove snakes could get drinking water from dew just as they get it from rain: allow the condensed water to trickle down their bodies to their mouths.

Mangrove snakes barely exceed three feet in length; the record is 93.3 cm. Most are notably smaller. Ashton and Ashton say they produce litters of up to eleven in the summer, and that like DeKay's snake some females may hold over fertile embryos for a spring brood. This seems not to have been proven.

Wright and Wright give litter size up to twenty-two, but this seems to be from hybrid individuals; the fresh water species get much larger and can have more young. The babies are six or seven inches long at birth (14 to 20 cm). There is some evidence that within a litter individuals of the color types found in salt water average smaller than those of the fresh water type (banded). The genetics and evolution of mangrove snakes and their fresh water relatives remain a great area for potential study.

Divergent color phases of mangrove snakes occur. Most are patterned in olive, grey, and brown. These usually show at least the beginnings of stripes on the neck, and may have a complete middorsal stripe. Other individuals are uniform orange, terra cotta, or brown. The scales are heavily keeled. The tail may be vertically compressed (as the species name suggests), but is not always.

These snakes eat mostly small fish and are great scavengers. Tide pools often provide stranded meals. They will also take frogs.

There are MCZ specimens from Key West, Stock Island, Summerland, Middle Torch, Big Pine, and Bahia Honda. I have examined and released live individuals on Little Torch, Big Torch, and Water Key north of Big Torch. I have records based on photographs from Archer, Cottrell, and Boca Grande—in the sand keys west of Key West. Duellman and Schwartz list additionally Little Pine, Lower Matecumbe, Key Vaca (Marathon), and Plantation Key. There is a museum specimen from the Marquesas (Carr, 1940).

References

Dunson, W. A. 1979. Occurrence of partially striped forms of the mangrove snake *Nerodia fasciata compressicauda* Kennicott and comments on the status of *N. f. taeniata* Cope. *Florida Scientist* 42(2): 102–12.

_____. 1980. The relation of sodium and water balance to survival in seawater of estuarine and freshwater races of the snakes *Nerodia fasciata, N. sipedon,* and *N. valida. Copeia* 1982(2): 268–80.

Lawson, R. 1987. Molecular studies of thamnophiine snakes: 1. The phylogeny of the genus *Nerodia. Journal of Herpetology* 21(2): 140–57.

Lazell, J. 1976. *This Broken Archipelago.* New York: Quadrangle, New York Times Book Co.

Malnate, E. V. 1960. Systematic division and evolution of the colubrid snake genus *Natrix,* with comments on the sub-family Natricinae. *Proceedings of the Academy of Natural Sciences of Philadelphia* 112: 41–71.

Miller, D. E. 1985. Rain water drinking by the mangrove water snake, *Nerodia fasciata compressicauda. Herpetological Review* 16(3): 71.

Rossman, D., and W. G. Eberle. 1977. Partition of the genus *Natrix,* with preliminary observations on evolutionary trends in natricine snakes. *Herpetologica* 33(1): 34–43.

BLIND SNAKE
(Typhlops braminus)

This species, one of the most widely distributed of all snakes, has obviously been transported by human agency.

Clifford Pope, 1935

This is a tiny species, one of the smallest of all snakes. A six-incher is big. I knew this species well and caught literally hundreds—including series of specimens on remote islands of the South China Sea—before I saw the first ever for the Florida Keys: MCZ 172619. It is a newcomer in our fauna, but I expect its spread to be explosive. It is known from the mainland in Dade and Palm Beach counties.

Pope describes it succinctly: "Small and worm-like; body covered above and below with uniform, cycloid scales arranged in twenty rows; eyes concealed under head shields; tail ending in a fine spine; 175 mm. or less in length." He gives its distribution as ". . . southern Asia and the Malay Archipelago including the Philippines. It also reaches the Pacific Islands, Africa, and Madagascar. . . ." Since Pope wrote, the species has also turned up in Mexico. Soon, no doubt, it will occupy all the warmer parts of the entire planet.

In China this "blind" snake (it can tell light from dark, but cannot see images) occurs north to the provinces of Fukien and Kiangsi, well outside

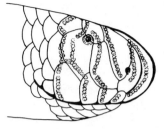

Head of the blind snake, *Typhlops braminus,* in side view. Scale edges are prominent. The eyes are less so, but visible. Dermal glands cluster along the sides of the scales.

the tropics. It attains elevations of 330 m in the Lofoshan of Guangdong, on the Tropic of Cancer. Thus, it can tolerate frost.

Females both lay eggs and produce live young. Clutches or litters vary from at least two to seven; eggs are said to measure about 4 x 13 mm. Although reproduction is said to be April to June, I believe they may produce young year 'round. Apparently no male *Typhlops braminus* has ever been found. This species seems to be truly parthenogenetic: a vast clone of similar females.

These snakes belong to a group called "scolecophidians"—"worm snakes." This one's family, Typhlopidae, has a few tiny teeth in the lower jaw, but none in the upper. A rather similar family, Leptotyphlopidae, has the opposite: teeth in the upper jaw, but none in the lower. Both families of worm snakes have remnant vestiges of pelvic girdles and hind limbs internally. It has been suggested that they are more closely allied to lizards than to true snakes.

I am not much on behavior as evidence of relationship, but I have observed a *Typhlops braminus* rattle! Many harmless snakes make a noise rather like a rattlesnake by merely vibrating their tails against a substrate. The *T. braminus* I observed vibrated so hard it shook the entire posterior third of its body. I have yet to hear of a lizard that rattled.

Typhlops are widespread and often common in the West Indies, but this is the only species yet reported in the United States. *Leptotyphlops* occur in Texas and surrounding states, but are very restricted in the West Indies, occurring there only on remote Providencia and San Andreas, several of the Lesser Antilles, and on Columbus's first landfall of San Salvador, or Watling, in the Bahamas.

Hawaiians are wont to say they have no snakes in their archipelago, but they are wrong. *Typhlops braminus* is egregiously abundant on all the major islands. Apparently the species was introduced first to Oahu in the early 1900s in potted plants from the Philippines. (Hawaii has a second, much larger, perfectly native species of snake too: *Pelamis platurus,* the yellow-bellied sea snake. It is hot in contention for the most lethally venomous animal on Earth, but you will be safe as long as you do not go in seawater.)

Males of some Australasian species of *Typhlops* have peculiar, near-embryonic hemipenes. This has led some snake doctors to classify them in a separate genus, *"Typhlina"* or *"Rhamphotyphlops."* Of course, with no males, one cannot place *Typhlops braminus* with this group for sure, although by inference it may be closely related to them. Since *Typhlops braminus* is the most abundant and widespread of all relevant species, one can see immediately the pragmatic nightmare generated by attempting to define genera on the basis of "males only" characteristics.

In September 1987, this species was known only from a small area in Old Town, Key West, in the Florida Keys. It will be most interesting to see how rapidly it spreads. It feeds mainly on small insect larvae, and may well prefer exotic species to native ones. We may never know if its presence in our islands is beneficial, or if it exterminates some nifty little species we never even knew.

References

McDowell, S. 1974. A catalogue of the snakes of New Guinea and the Solomons, with special reference to those in the Bernice P. Bishop Museum. I. Scolecophidia. *Journal of Herpetology* 8: 1–57.

Pope, C. H. 1935. *The Reptiles of China.* New York: American Museum of Natural History.

Wynn, A., C. Cole, and A. Gardner. 1987. Apparent triploidy in the unisexual Brahminy blind snake, *Rhamphotyphlops braminus. American Museum Novitates* 2868: 1–7.

REEF GECKO
(SPHAERODACTYLUS NOTATUS NOTATUS)

It is found commonly in the woods in Cuba and the Bahamas. . . .
When collecting insects in the forest, and tearing open rotten logs,
one often meets with small single eggs, pure white, and about 3 x 5
mm. in size, and as the obvious parent is so often near by there
cannot be much doubt but that these are the eggs of
Sphaerodactylus.

BARBOUR AND RAMSDEN, 1919

These tiny lizards are even more common in the Florida Keys than I have found them in either Cuba or the Bahamas, and our Keys are where they were first named—by Spencer Fullerton Baird, who wrote railroad survey reports and, in so doing, may have earned the title of America's foremost mammalogist. I cannot imagine how Baird managed to find and describe our diminutive and endemic form. One day I will look up the whole story.

Meantime, these are the minuscule lizards of trash piles, beach debris, and leaf litter in the hammocks and pinelands most any curious naturalist will uncover. Geographic variation in Baird's little lizards has been thoroughly studied by Albert Schwartz (1966). A widespread subspecies is in Cuba, the Isle of Pines, the Morant Cays south of Jamaica, and on Great Inagua in the Bahamas. Another occurs on the remote Swan Islands far west of Jamaica. Two subspecies occupy the two major Banks of the Bahamas, respectively. The nominate form of the species occurs only in Florida, from the Tortugas to Broward County.

Smith (1946) says Collier County too, which he must have got from Carr (1940). I can find no reference to a Collier County specimen. Schwartz says only Monroe, Dade, and Broward. There are museum specimens from the Tortugas (Garden Key), Key West, Stock Island, Cudjoe, Summerland, Middle and Little Torch, Big Pine, Upper and Lower Matecumbe, Plantation, and Key Largo. Duellman and Schwartz add Teatable and Carr adds Lignumvitae, but I do not know if these are represented by actual specimens. I have examined and refrained from collecting specimens on Cudjoe Key, too.

In the 100 m² cover turn-out on two hectares at Snake Acres in hammock L9/16, Middle Torch Key, we average about five each time. I have found as many as that under one square meter of cover. I cannot judge an activity period for these lizards. I have occasionally seen one out in the open both day and night. They are generally active whenever uncovered.

The first Tortugas specimens seem to have been collected in the summer of 1938 by the Van Hyning-Russell team (Carr, 1940). They were deposited in the Florida State Museum. Since then, the species has been collected several more times, at least on Garden Key. Schwartz (1966) had one in the University of Michigan Museum of Zoology. Bill Robertson has seen them; no one doubts they are really there.

Dr. Holder, James Mills, Captain Harrison, and other nineteenth-century collectors did not secure them. Why? Were the collectors inept? Were the lizards not present? This species disperses readily and might easily travel—as lizards or eggs—in freight like lumber or ornamental plants. That is the suggested method for the Cuban subspecies' presence on Great Inagua and the Morant Cays. On the other hand, the Florida populations must be native: they are widespread but homogeneous in characters, and very well differentiated. They may well have got here in the late Sangamon. They could and should have been stranded on the Tortugas by post-Wurm sea level rise.

For tiny lizards—usually about an inch (2.5 cm) head-body, with a shorter tail—these fellows have big scales. In adult males some scattered scales are dark, making a spotted brown animal. Females tend to have real patterns. The dark brown coalesces into stripes, especially on the head. There is usually a sooty "target" on the shoulders containing a pair of near-white spots. Carr (1940) says they lay their eggs in "June to August, singly or by twos or threes in rock piles and under boards; they are oval, and average 6 by 4.5 mm. The young are 23–25 mm. long and like the adult in pattern and coloration."

This makes sense because when I work the cover from December to March I get adults, never eggs. I have found hatchlings, which looked like adult *females,* in August, and eggs *inside* thatch palm logs. These geckos, like all their relatives, are insectivorous.

References

Barbour, T., and C. T. Ramsden. 1919. The herpetology of Cuba. *Memoirs of the Museum of Comparative Zoology* 47: 69–213.

Schwartz, A. 1966 ("1965"). Geographic variation in *Sphaerodactylus notatus* Baird. *Revista de Biologia Tropical* 13(2): 161–85.

ELEGANT GECKO
(SPHAERODACTYLUS ELEGANS)

Chiefly crepuscular, but active at night near house and street lights; hides in crevices or under debris or vegetation during daylight hours. Often found in or on walls of cisterns and outhouses.

ROGER CONANT, 1975

Outhouses indeed! At Snake Acres there is a castle on whose walls these lovely, small lizards crawl. I have bales of notes on their activity and behavior from sunset to after 1:00 a.m. in the morning (the latter a very unusual night for me). Admittedly, they were in the trailer, on walls or screens, under lighted conditions, but not close to the lights. Once at 10:45 p.m. I entered the totally dark trailer, switched on a light, and there was a fine adult out and walking across the wall. So they can be quite nocturnal.

I have found them standardly at about two meters from the ground, often at three meters, and once—active at 11:00 a.m. on a screen—at four meters on a second story (Cudjoe Key). In 22 100 m² turn-outs of cover on two hectares in hammock L9/16, Middle Torch Key, we have seen only three individuals, total, of this species. It is far more of a house gecko than is *S. notatus* or most other *Sphaerodactylus*.

I grew up calling this species *Sphaerodactylus cinereus*, the "ashy gecko." Normally I resist name changes, but I was delighted to learn that the name given to the juvenile had priority over *cinereus*. The first-given, now official, name is *elegans*. And elegant they are. Hatchlings are gaudily banded with black on a yellow or olive-grey ground color, shading abruptly to brilliant red on the legs and tail. Adults are lemon-yellow with a fine reticulation of slate-grey. Adults at a distance look fairly drab, because their markings are so fine, but just get close up. These geckos move with grace and stealth, seeming to almost glide across a wall. When they encounter each other they are apt to lash their tails in slow motion. You have to see it to appreciate this action.

In all the standard literature this species is recorded from Key West (and sometimes Key Largo: Smith, 1946). Duellman and Schwartz added Boca Chica. Carr (1940) extended the range to Dade County, but cites no specimens or localities. I have seen museum specimens from Key West, Summerland, Stock Island, Middle and Little Torch Keys. I have seen living individuals on Cudjoe, Big Pine, and remote, uninhabited keys like Spotswood off Ramrod. This species is widespread in Cuba and Haiti, where I first got to know it as a callow lad.

It is conventional wisdom that it was introduced recently to the Keys, presumably from Cuba. Forty-five years ago Carr (1940) pointed out: "However, inquiries among the most patriarchal 'conchs' I have been able to locate have convinced me that all . . . lived in Key West long before they were ever discovered by herpetologists." (He included *Hyla septentrionalis* and *Eleutherodactylus planirostris* too.) No evidence suggests that *S. elegans* is not a native of the Lower Keys, perhaps made more abundant by the erection of edifices to its liking.

Adults are about three inches long, with the tail. The head-body record is 3.7 cm—much larger than *S. notatus*. The eggs, Smith (1946) says, "like tiny bird eggs," are said to be laid in summer. Juveniles are easily found in September, but scarce in December. I have, in the winter months, found specimens changing their colors—intermediate between juvenile and adult patterns. I deposited one, in a series caught on Little Torch Key, in the Australian Museum, Sydney. Examples from our Keys are also now in the Chengdu Institute of Zoology, Sichuan, China.

OCELLATED GECKO
(Sphaerodactylus argus argus)

Recent collecting has shown no evidence of this species.

Duellman and Schwartz, 1958

This is a tiny, brown, ground-dwelling lizard rather like *S. notatus,* except that it has numerous white spots on the nape and shoulder region and much smaller scales. Although certainly present in Key West prior to and up till 1944, it was not known to Carr (1940) or Smith (1946). Duellman and Schwartz (1958) saw three specimens. Conant in 1958 did not include it, but in his 1975 edition he provides a detailed description and the comment: "The tiny white eyelike spots (ocelli) on the nape are far fewer than the hundred eyes of Argus of mythology, but they give this lizard its scientific name and furnish a quick clue to identification."

Adults get to about two-and-a-half inches (6 cm) total length, and a maximum of 3.3 cm head-body length. I know nothing of its reproduction, but presumably it is quite like *S. notatus.*

This species is certainly a native of Jamaica, the only place I have ever encountered it. It seems clearly to have stowed away to Bimini, Nassau, and Key West. It also occurs on the Corn Islands off Nicaragua, in southern Cuba, and the Jardines de la Reina.

If you know anything about this species, or know where I can find one in the Keys, I would surely like to hear from you.

MEDITERRANEAN GECKO
(HEMIDACTYLUS TURCICUS TURCICUS)

The speed with which this gecko has expanded its range in the southern U.S. in recent years is even more remarkable than the spread of the cattle egret. The bird has wings. . . .

ROGER CONANT, 1975

Hobart Smith (1946) gives its original range as "Sind and Persia west to the borders of the Red Sea; Socotra Island and . . . Somaliland . . . north to Egypt and west around the Mediterranean Basin to Morocco and the Canary Islands." Fair' makes my feet itch.

These are big geckos, up to five inches total length (head-body maximum 6 cm), with grand, expanded, leafy toes and warty hides. They are true homebodies, but your home will do; they will not build their own.

Like most geckos, the large eyes are covered with a solid, immobile spectacle like snakes. Their pupils are elliptical. Their colors are shades of brown and grey, highly changeable according to mood and time of day, and usually boldly blotched. Like most of the larger geckos, these have a voice: "a faint mouselike squeak" (Conant, 1975).

Frankenberg's studies of these geckos in their South Florida range provide fascinating details: they have a language. Baby Mediterranean geckos make three recognizably different squeaks, all related to escape. Adult females make four sorts of calls signaling approval, defensiveness, threat, and desire for release. The latter, used to notify amorous males when the female is just not in the mood, is unique. Adult males have the most complex language. They make the same approval, defense, and threat calls as females, but lack a release call and add advertisement and attack calls. The advertisement call consists of clicks and changes as soon as the male recognizes the sex of the other gecko to whom he is advertising his presence.

When a male Mediterranean gecko perceives that another individual in sight is also a male, his advertisement clicks become largely irregular in spacing. This will never lead to an approval call, but instead either a defense or threat call. Unless one male or the other backs down, an attack will likely be sounded. Usually, in natural situations, the males hold well-spaced territories and actual attacks are infrequent.

When a male perceives that the other gecko in view is a female, however, his advertisement clicks become regular. He no doubt hopes this will lead to an approval call from the female.

Frankenberg compared Mediterranean and Indo-Pacific geckos (the next species) ecologically and behaviorally too. They are quite similar in size, diet, and activity patterns. However, they differed in perch preference, patterns of site specificity, and—most notably—in social organization. Essentially, Mediterranean geckos are rather social creatures, as their complex (for a lizard) language indicates. They tend to be more often out in the open and visible and are relative stay-at-homes.

Indo-Pacific geckos wander more and get in more fights. They win those fights more often, too. Frankenberg found Mediterranean geckos much scarcer than Indo-Pacifics on Key Biscayne: the ratio was about one to five. On Key Vaca, however, Mediterraneans outnumbered Indo-Pacifics nearly seven to one.

A classic stowaway in lumber and crates, this species has been in Key West and Miami since before World War II (Carr, 1940). Duellman and Schwartz also saw it from Big Pine Key, where Frankenberg found it too. William Dunson deposited a specimen in MCZ from Summerland Key, where Frankenberg found only Indo-Pacific geckos. I have not seen it alive in the Keys, and Carr (1940) did not find it common anywhere.

Smith (1946) notes: "In tropical America, where they are common in and near port towns, they are usually dreaded as poisonous creatures. Actually they are quite harmless and delicate." He found this species was especially fond of ants and termites. In studying a close relative (H. mabouia) in the Virgin Islands, we have found the plurality of objects in the diet to be roaches. In any case, they are nocturnal, love edifices, and are extremely beneficial to their human commensals.

I can find little on reproduction or life history of this species in America. Carr (1940) reports a pair of eggs in mid-June under a board in a stable, but he does not say whether in Key West or Miami. One reference provided below, albeit in German, will give you what is known about the species in its native Mediterranean region.

References

Frankenberg, E. 1982. Vocal behavior of the Mediterranean house gecko, *Hemidactylus turcicus. Copeia* 1982(4): 770–75.

_____. 1984. Interactions between two species of colonizing house geckos, *Hemidactylus turcicus* and *Hemidactylus garnoti. Journal of Herpetology* 18(1): 1–7.

Salvador, A. 1981. *Hemidactylus turcicus* (Linnaeus 1758)—Europaischer Halbfingergecko. *Handbuch der Reptilien und Amphibien Europas*, Akademische Verlagsgesellschaft, Weisband 1(1): 84–107.

INDO-PACIFIC GECKO
(HEMIDACTYLUS GARNOTI)

The Indo-Pacific Gecko, sometimes called the Fox Gecko because of its long narrow "fox-like" snout, is a member of a tropicopolitan genus renowned for its ability to live in association with man.

SEAN MCKEOWN, 1978

This is a newcomer to Florida and has not yet apparently established itself in the Antilles or Bahamas. Clifford Pope notes its type locality is Tahiti and that it occurs in tropical China and on Hainan Island. He gives its range outside of China: "Widely but sporadically distributed from Sikkim and Burma southeastward through the peninsula of southeast Asia and the East Indies to islands of the South Seas." How can you stand to just sit there and take his word for it?

I couldn't; I admit I have not gotten to all the places included in that sweep, but I have been to many, and *Hemidactylus garnoti* is usually there. I know it well in the field. I know it well enough right at Snake Acres on Middle Torch Key.

The first time it appears in the general works cited above is in Conant (1975). He records it from Miami and Sanibel Island. Smith and Brodie (1982) only give Miami. Steiner and McLamb (1982) collected it at Garden Key in the far Tortugas in September of 1981. William Dunson got three on Summerland Key in 1979; these are now in MCZ, Harvard, along with a voucher from my place on Middle Torch. Frankenberg, cited above under Mediterranean gecko, got this species on Key Vaca and Key Biscayne, as well as Summerland. I wonder how it reached Florida but missed the rest of the Atlantic-Gulf region?

As noted in the preceding account, this species is similar in size and build to *Hemidactylus turcicus* but lacks the big, wartlike scales scattered around the body and has more lamellae—big, overlapping scales—under its leaflike toes. It also has a saw-tooth look to the tail caused by an enlarged, pointed scale at each side of each caudal annulus (ring of tail scales): best seen by looking at the underside of the tail. It is less boldly marked than *H. turcicus*, but just as changeable in color from light to dark shades of grey and brown. It is usually lemon-yellow on the belly and orange or pinkish under the tail. A big individual will total more than five inches (13.3 cm), but they rarely have complete, original tails. Head-body maximum is 6.4 cm.

These lizards are said to be wholly parthenogenetic: a clone of females producing more females asexually. This certainly helps them spread around. It only takes one to make a colony. This parthenogenesis is thought to work because potential eggs—oocyctes—undergo a chromosomal duplication without cell division prior to entering *meiosis*—the reduction division that, in normal animals, makes monoploid eggs or sperm. The penalty for this method of reproduction is lack of recombina-

Indo-Pacific gecko, *Hemidactylus garnoti.*

Louis Porras

tion, which gives variety and diversity to the gene pool. If you are already perfect, you do not need to be more variable. Who says there is little to be learned from lizards?

In Hawaii, Hong Kong, and Hainan these geckos lay hard-shelled, nearly round, white eggs about the size of double-oh buckshot. Each lizard and her clone mates (sisters?) may return repeatedly to the same laying site: under a sill, behind a shutter, in a desk drawer, or (how mundane) beneath a log. The eggs stick to the surface laid on; there may be rows of them in various stages from just laid to hatched. I detect no seasonality to their reproduction in the tropics.

In Miami Voss found Indo-Pacific geckos lay eggs from June to January. They each lay two eggs at a time. The eggs averaged 7 to 10 mm in diameter and hatched in about sixty days. Hatchlings measured 24 to 26 mm snout-vent length (abbreviated SVL; does not include the fragile tail).

In the South Pacific, Gibbons and Zug reported on a single egg measuring 9 x 10 mm that weighed 0.6 gram (about two one-hundredths of an ounce). They measured three hatchlings at 27 and 28 mm (average 27.3) SVL; these had 19 or 20 mm tails and weighed 28 to 36 grams (average 31): from one ounce to slightly more.

Their diet seems to be quite like other *Hemidactylus:* all insects, lots of roaches. They are not unwelcome at my house.

References

Gibbons, J. R., and G. R. Zug. 1987. *Gehyra, Hemidactylus,* and *Nactus* (Pacific geckos). Eggs and hatchlings. *Herpetological Review* 18(2): 35–36.

McKeown, S. 1978. *Hawaiian Reptiles and Amphibians.* Honolulu: Oriental Publishing.

Pope, C. 1935. *The Reptiles of China.* New York: American Museum of Natural History.

Steiner, T. M., and L. T. McLamb. 1982. Geographic distribution: *Hemidactylus garnoti. Herpetological Review* 13(1): 25.

Voss, R. 1985. Notes on the introduced gecko *Hemidactylus garnoti* in south Florida. *Florida Scientist* 38(3): 174.

YELLOW-HEADED GECKO
(GONATODES ALBOGULARIS FUSCUS)

A thriving colony was found April 23, 1939, on the dredged-up land occupied by the old railroad docks on the western side of Key West; this is the area known as Trumbo. . . .

ARCHIE CARR, 1940

This is the most ungecko-like gecko we have. In obvious features it seems quite intermediate between geckos and anoles, the next group. The toes lack the spherical pads or leafy expansions of our other geckos. There are rudimentary eyelids. Except in bright light, the pupils appear nearly round. The sexes are strikingly different; the males brightly colored. They are diurnal, even though they do like shady places.

I lack field familiarity with our subspecies, but I know the Haitian form well. It is very brightly colored; ours are less so. The name *albogularis* means white-throated, probably based on a faded alcoholic specimen; the subspecies name *fuscus* just means brown. In life, the adult males of our form are grey, changing from nearly black to bluish, with a yellow or orange-yellow head and neck, often somewhat mottled with grey or brown. The females and juveniles begin life with brown bands and grow to be variably mottled. An ashy-whitish, dark-bordered neck line or collar is a striking marking. The original tail tip is pallid; it looks bleached or sun-faded.

These are small lizards, but not tiny. They attain about three-and-a-half inches (9 cm) total length; the head-body maximum is 4 cm. I know of no reproductive data from the Keys, but elsewhere they seem rather like *Hemidactylus*. They often lay eggs in a special spot, where many can accumulate, although probably laid one at a time. The eggs are hard-shelled and white, about 8 mm by 6.5 mm. I doubt there is much seasonality to reproduction in the tropics, but there may be here.

I do not know the original range of this stowaway species of strongly edificarian habits. It is found in widely scattered parts of Cuba, Jamaica, a few more remote islands, and central and northwestern South America. It is also established in Coconut Grove. In the Keys it is known only from Key West.

Smith (1946) cites Major Chapman Grant's (1940) account of this form in Jamaica as his authority. Fortunately, this now rather rare book is before me (thanks to some of my youth not misspent). Grant, in turn, quotes a color description of Cuban individuals. It says the males do have white throats in life, and often red and blue markings on the head. Grant's description of a Jamaican male agrees with my views of another subspecies in Haiti: throat orange, like the rest of the head. The eggs do not stick to their substrate like *Hemidactylus* eggs.

Reference

Grant, C. 1940. *The Herpetology of Jamaica II. The Reptiles*. Kingston: Institute of Jamaica.

GREEN ANOLE
(ANOLIS CAROLINENSIS)

Although the genus Anolis *Daudin, 1803, is the largest of all American lizards, containing some 250–300 species, its type was apparently never dealt with. . . . Our own work demands attention to the problem which surely requires something less than a "Ph.D. thesis for a nomenclaturist" . . . for arrival at a definitive disposition acceptable to all students involved.*

SMITH, WILLIAMS, AND LAZELL, 1963

The genus *Anolis* seems to be the largest valid genus of amniote vertebrate animals. There have been many attempts to split it, and an occasional splinter may be shaved off, but this great assemblage, found throughout the warmer parts of the New World, stands monolithic. Recent attempts to resurrect, for example, *"Norops"* for some of its species are ill-advised to the point of being silly.

When we petitioned the International Commission to make *Anolis carolinensis* the official type species of the genus, we believed we had a solid case everyone would accept. A genus needs a type species so that workers can have a sort of touchstone: something to attach the system of systematic biology to. I have alluded to this problem under *Natrix* (mangrove snake), above. We hope this now exists for huge *Anolis*.

Anolis carolinensis is a problematical entity. It is closely related to several Cuban and Bahamian forms, some of which are accorded full species rank even though no one seems quite sure how to distinguish them. There are certainly two full species in Cuba. They are easy to tell apart in the middle of that island, where they both occur together. Which is closer to *Anolis carolinensis* is a moot point.

Our species, a gorgeous green lizard patterned with blue and cream, sporting a bright pink throat fan, is mightily variable. For a start, any individual can change color right before your eyes. This feat earns them the name "chameleon"—despite the fact that they do not change to match their backgrounds and are no relation to true chameleons of the Old World. Ours often turn grey-brown, or even rusty, with a prominent, dark-bordered ashy stripe down the back (especially in females and juveniles). The blue pattern on the shoulder and rib cage region becomes ash-grey in this coloration. When the animal is upset, the area behind the eye turns sooty blackish.

There is also great geographic variation. Ours have elongate, rather spatulate heads; those from farther north, into the Carolinas, have shorter heads. Those from Texas have the shortest heads of all and are rather dull-colored with a double row of white dots on each side. A striking form from the Fort Myers vicinity has a blue or blue-grey throat fan. Populations on the Gulf Coast barrier islands may be extremely pale and have deep purple fans. No one has attempted a taxonomic resolution of all this complexity.

Anoles have a pad of widened, overlapping scales under their toes (except the terminal phalange). This is reminiscent of the toes of *Hemidactylus* geckos except the *Anolis* scales—properly called lamellae—are not divided. Males get almost twice the size of females, reaching about eight inches (20 cm) total length. The head-body maximum is 7.5 cm. Females usually lay one egg at a time, and tuck it under debris or leaf litter on the ground.

Anoles are sort of the wood warblers of the reptile world. They are active by day, often conspicuous, and sport bright colors. Hobart Smith (1946) writes: "These little creatures are one of the most interesting features of the American tropics and of southeastern United States. Their active clambering about trees and fences, their flashing fans and dashing combats, their breath-taking falls and indignant recoveries form an endless and constantly amusing repertoire which one can watch with interest indefinitely."

In natural habitats these anoles are not especially common. They are widespread, however, and do well on remote keys that have a bit of hammock or even transition buttonwoods. They are more common in gardens. I have seen museum specimens from Key West, Stock Island, Summerland, Middle Torch, Big Pine, No Name, Indian Key, and Upper Matecumbe. I have color and habitat notes on living individuals additionally from Boca Grande (there is also a museum specimen: FSM 76327), Upper Sugarloaf, Cudjoe, Little Pine, and Lignumvitae. Carr (1940) notes additional museum specimens at FSM from two of the

Marquesas (details clarified by Dodd and Possardt, 1987) and Woman Key in the outer keys. Duellman and Schwartz add Boca Chica, Long Key, Lower Matecumbe, Plantation Key, and the Newfound Harbor Keys off southern Big Pine. They also list "Sand Key," possibly Sandy Key in Florida Bay (not the one southwest of Key West, but maybe), and Ragged and Totten's Keys, which are north of Key Largo.

Two other species of closely related green anoles have been introduced to the Florida Keys, both from Cuba. Allen and Slatten, back in 1945, reported two specimens of *Anolis porcatus* among 17 *carolinensis* collected at Key West. Their specimens were identified by the late Karl P. Schmidt, who certainly seems to have known how to tell the species of green anoles apart. If *porcatus* and *carolinensis* really do occur (or even occasionally have occurred) together without interbreeding, this would go a long way toward resolving thorny taxonomic issues. Typical *carolinensis* and Cuban *porcatus* are easily distinguished on head shape and scale counts, but Lower Keys *carolinensis* seem to at least approach the *porcatus* condition. Accepting such an approach as evidence that we are seeing geographic variation in a single species (uniting *carolinensis* and *porcatus* as subspecies) would be contradicted by independent occurrence in sympatry—even temporarily. Those old specimens should be found and critically compared to other Keys and Cuban material.

The second interloper is the Cuban giant, or knight, anole: *Anolis equestris*. Bright green with cream to rich yellow on the head and as a stripe over the forearm, these spectacular and stealthy creatures may reach nearly 20 inches total length (snout to vent about seven inches, or 17.9 cm)—the size of a young iguana. Larry Brown found this species on Elliott Key north of Key Largo. It is well-established in the Miami area. Conant (1975) provides a good account of it.

References

Allen, E. R., and R. Slatten. 1945. A herpetological collection from the vicinity of Key West, Florida. *Herpetologica* 3(1): 25–26.

Brown, L. N. 1972. Presence of the knight anole (*Anolis equestris*) on Elliott Key, Florida. *Florida Naturalist* 45(4): 130.

Dodd, C. K., and E. E. Possardt. 1987. Geographic distribution: *Anolis carolinensis* (green anole). *Herpetological Review* 18(1): 20.

Smith, H. M., E. E. Williams, and J. Lazell. 1963. *Anolis* Daudin, 1803 (Reptilia; Lacertillia): request for the designation of a type species under the plenary powers. *Bulletin of Zoological Nomenclature* 20(6): 438–39.

Williams, E. E. 1983. Ecomorphs, faunas, island size, and diverse end points in island radiations of *Anolis*. In *Lizard Ecology*, R. B. Huey, et al., eds. Cambridge, Mass.: Harvard University Press, pp. 326–70.

BROWN ANOLE
(ANOLIS SAGREI)

Dear Barbour:
  The Anolis arrived a few minutes ago—I have waited for it forty
years! for I have the female to match your male! . . .

LEONARD STEJNEGER (IN BARBOUR, 1931)

This is by far the most abundant and conspicuous lizard in the Florida
Keys. It was not always so. Carr (1940) remarked on how concentrated it
was in the northwestern part of Key West and how it was not generally
distributed around the island. Smith (1946) knew of it only from Key
West. Duellman and Schwartz added Cudjoe Key. Conant in 1958 said,
"Key West and nearby Keys," but by 1975 had one sort or another of *Anolis
sagrei* all over South Florida, north to Palm Beach and St. Petersburg.
Weingarner et al. (1984) first documented it in the Dry Tortugas on
Garden Key. Steiner and McLamb (1985) found it on Loggerhead Key, the
present limit of land westward in the Tortugas.

Julian Lee, University of Miami, is in process of studying variation in all
these populations, thought to have been severally derived from Cuba and
the Bahamas, and belonging to at least two subspecies. Lee's papers
meticulously detail scale count variation, size, and proportions, but leave
out color. Scale characters and size seem not to elucidate the taxonomic
questions.

These are stockier, shorter-headed anoles than the green anole. They
are always olive or brown, but can turn light or dark. In the Keys they have
rich strawberry *red* throat fans with near-*white* scales and a *yellow* border.
Nominate *A. s. sagrei* that I have seen from the Tampa-St. Pete region
have dull orange to olive throat fans with an ashy-white border. The
Bahamian form, introduced into Miami, is said to have an orange to grey—
or grey spotted—fan, with the whitish edge dark spotted too. The "*white
streak* down center of throat" noted by Conant (1975) fits the certainly
introduced mainlanders in Florida, but, because that streak is really
the throat fan border, does not fit Keys specimens. In the Keys it is
yellow.

Because of the three seemingly quite different sorts of throat fans
(smaller, but very much present in females), I do not know how to allocate
subspecies names. Possibly Thomas Barbour's name "*stejnegeri*" may yet
stand for at least Lower Keys populations and presumably for some Cuban

populations too. I do not doubt the species was introduced to the Keys in the rather recent past. I just wonder from whence it came. Stejneger's evidence, in Barbour's paper, indicates the species has been in Key West since at least 1890.

These anoles can exceed eight inches total length (21 cm), but most of that is tail. The maximum head-body length is 6.4 cm. Males are much bigger than females. Females and young have the boldest yellowish mid-dorsal stripes, but these can be evident in adult males too.

These anoles do not climb as high as green anoles and move more rapidly and erratically. Like most anoles so far studied, they tend to lay one egg at a time, about every two weeks, pretty much year 'round unless the weather is very cool or dry. The eggs are laid in leaf litter or under loose objects, often at the base of a vertical perch like a sapling. This habit sequesters the eggs effectively in potted plants and nursery stock, and facilitates the rapid spread of the species.

I have examined museum specimens from Key West, Stock Island, Summerland, Middle and Little Torch, No Name, and Bahia Honda. Lee adds Key Vaca, Key Largo, Plantation, Grassy, Big Pine, Long, and Upper Matecumbe. I have notes on individuals on Cudjoe Key and bales of behavioral notes from Snake Acres on Middle Torch. About the only point of immediate interest in all of that is that *Anolis sagrei* may become quite nocturnal around a light that is regularly left on at night.

References

Barbour, T. 1931. A new North American lizard. *Copeia* 1931(3): 87–89.

Lee, J. C. 1985. *Anolis sagrei* in Florida: phenetics of a colonizing species. I. Meristic characters. *Copeia* 1985(1): 182–94.

———. 1987. *Anolis sagrei* in Florida: phenetics of a colonizing species. II. Morphometric characters. *Copeia* 1987(2): 458–69.

Steiner, T. M., and L. T. McLamb. 1985. Geographic distribution: *Anolis sagrei*. *Herpetological Review* 16(4): 115.

Tokarz, R. R. 1988. Copulatory behaviour of the lizard *Anolis sagrei*: alternation of hemipenis use. *Animal Behaviour* 36(5): 1518–24.

Weingarner, C., W. Robertson, and W. Hoffman. 1984. Geographic distribution: *Anolis sagrei*. *Herpetological Review* 15(3): 77–78.

BARK ANOLE
(ANOLIS DISTICHUS DOMINICENSIS)

Resembles a piece of lichen-covered bark that scurries away when approached.

ROGER CONANT, 1975

This is our smallest and scarcest anole. It attains about five inches (over 12 cm) total length, with a head-body maximum of 5.1 cm. The males and females are not strikingly different, although males do grow larger. The colors are lichenate shades of grey-green and olive; the throat fan—well-developed only in the male—is pale yellow with an orange central suffusion. Reproduction is probably like the other anole species, but I am not certain.

One subspecies in the Miami area, *A. d. floridanus,* although named first in Florida probably originated in the Bahamas. Our subspecies has also been introduced in Miami. In the Keys it occurs to date only in Key West, where I first found it 12 June 1977. I deposited specimens in both MCZ, Harvard, and FSM, University of Florida. I found the species common on the big trees in the trailer park at Simonton and Virginia Streets. It was still common there in May 1985. Someone might get a fine paper and recognition out of studying this population. If it takes off as *A. sagrei* has done, it could be elucidating to study the process. As yet we know little in specific terms about colonizing species. One should compare directly to Lee's work on *Anolis sagrei.* In addition, I provide a grand classical reference to whet your appetite for the task.

Dr. F. Wayne King wrote his doctoral dissertation on the ecological relationships of this species and *Anolis carolinensis* in Miami. He found that while the green anole, *carolinensis,* usually stalks its prey stealthily, bark anoles, *distichus,* usually make a running attack. Probably this difference accounts for the fact that green anoles eat mostly dipteran insects, like flies, while bark anoles eat mostly ants. Green anoles also tend to prefer open, sunny sites, while bark anoles prefer shady sites.

References

Elton, C. 1958. *The Ecology of Invasions by Animals and Plants.* London: Methuen Co.

King, F. W. 1966. Competition between two South Florida lizards of the genus *Anolis.* Ph.D. dissertation, University of Miami, Coral Gables.

FIVE-LINED SKINK
(Eumeces inexpectatus)

The natural history of this species is well worth investigation, as an aid to a comparative study of the three five-lined species.

HOBART SMITH, 1946

The late E. H. Taylor, one of our greatest herpetologists, first recognized and described this species in 1932. At that time, discovery of new species was commonplace, even in the United States. Why Dr. Taylor chose the name he did—the "unexpected" skink—is a mystery to me. Nowhere in the original description does he allude to it.

Although there were at the time of Taylor's description specimens from the Keys, notably Dr. Holder's two from the Tortugas, all of the material Taylor examined was from north and central Florida. Smith (1946) called this the "Floridian" five-lined skink, although he knew it occurred west to Louisiana and north to Virginia. The current official common name is "southeastern" five-lined skink. All three species of five-lined skinks occur in all of the southeastern states. Ours is the only one of them that lacks a conspicuously enlarged line of median scales under the tail. All of its tail scales are about the same size. Skinks are *shiny* lizards.

Young specimens are boldly striped with cream or yellow on a brown or blackish ground and have brilliant blue or purple tails. Females retain this pattern throughout life, but the median stripe—always narrow—may become obscure and the tail blue more subdued. Red suffusions are present on the head. Old males become virtually bronze-brown all over with orange heads.

Our species may attain about eight-and-a-half inches (21 cm) total length. The maximum recorded head-body length is 8.9 cm. The eggs, laid in spring or early summer (data from well north of the Keys), are about 13 mm by 7 mm, white, and leathery. They hatch in a month or two. The hatchlings are about 1 inch (25 mm) long, head-body, with a longer tail.

I regard this species as scarce in the Keys. I have encountered it twice at Snake Acres on Middle Torch Key; that works out to about once every six or seven years. Three specimens in MCZ, Harvard, are from Key West and two from the Tortugas. A specimen is at FSM from Boca Grande in the outer Keys. An MCZ specimen labeled "Pine Key" may have come from Big Pine (or Little Pine, or Long Pine up in Everglades). Duellman and Schwartz give a good roster, adding to my list Big Pine (for sure), Little Pine, Boca Chica, Cudjoe, Sugarloaf, Indian Key, and Key Largo.

Although these seemingly brightly colored lizards bask in the sun and climb on trees and buildings, it is possible for them to live in an area for years and not be noticed. My experience with them at my own place on Middle Torch makes the veracity of Holder's Tortugas records all the more plausible.

Reference

Taylor, E. H. 1932. *Eumeces inexpectatus:* a new American lizard of the family Scincidae. *Kansas University Science Bulletin* 20(2): 251–71.

KEY MOLE SKINK
(EUMECES EGREGIUS EGREGIUS)

The Florida Keys Mole Skink is found in sandy areas, usually near the shoreline, but not always. . . . It probably requires a fairly loose soil in which it burrows in the characteristic "sandswimming" manner.

STEVEN P. CHRISTMAN, 1978 (IN MCDIARMID)

This is an endemic form, confined to the Florida Keys. It is our only state-listed species of lizard, regarded as threatened (Christman in McDiarmid, 1978) by development. I regard it as genuinely rare and believe it is probably endangered. Carr (1940) found it "locally common" but I have not found such a locale. The type locality is Indian Key, where I have searched but failed to find it. The distinctive characteristics—lateral stripes continuous to groin and middorsal scale rows enlarged—seem best developed in the Lower Keys.

Anywhere in the Keys, this is the only small, slender, red-tailed lizard. A big one is six inches (15 cm) total length; most of that is the extremely fragile tail. The head-body maximum is 5.7 cm. Males and females seem so similar I cannot separate them externally. The tails of those I have seen were bright red, but the books say reddish brown to pink. In an excellent short account, Conant (1975) says the males develop "a reddish or orange suffusion that extends onto the venter during the mating season," but he does not say just when that is. Smith (1946) says: "Practically nothing is published on the natural history of these interesting skinks." More than three decades later Christman (in McDiarmid, 1978) says, "Little is known about the life history and ecology of this animal."

L. H. Babbitt recorded courtship and mating of these skinks at Key West, witnessed on 13 March 1946. For 17 of 25 minutes, "courtship" consisted of the male literally dragging the female around. He spent two minutes chasing her, five minutes actually copulating, and took a one-minute rest during his dragging time. One might anthropomorphize that this sort of behavior could lead to rarity in a species.

In a series of papers Dr. Robert H. Mount of Auburn University, Alabama, has elucidated the taxonomy of the species as a whole and aspects of the life history of mainland forms. These and other references are given in his summary paper of 1968. Florida mainland red-tailed or mole skinks (some subspecies do not have red tails) mate in March. The females lay three to seven eggs in a burrow in sand or loam and remain with the eggs until they hatch in 31 to 51 days. The young skinks attain sexual maturity in about a year, at least under laboratory conditions.

Great caution must be exercised in applying these data to our mole skink. As Christman points out, ours seems not to be gregarious (I have only ever found singles) while the mainlanders are strikingly so. Of course the pronounced seasonality of the mainland is mitigated in the Keys.

I have found these animals difficult to handle because they twist and break off their tails—sometimes in several pieces—even when I carefully avoid touching that member. In all my searches under good cover on Middle Torch Key (see account of ringneck snake, above) I have found this animal just once. A specimen from Key Vaca is in the Australian Museum, Sydney. MCZ has specimens from the Tortugas and Key West. Duellman and Schwartz record it additionally from Indian Key, Stock Island, Upper Matecumbe, and Key Largo. Specimens from the latter are intergradient between the Key form and the mainland subspecies *onocrepis* (see references in Mount). Christman (in McDiarmid, 1978) provides a photograph of one from the Saddlebunch.

Carr (1940) found these skinks among rocks close to the sea, where they may vary the usual insect diet of lizards by eating small crustaceans.

In view of the general rarity of this lizard, I cannot doubt the veracity of Dr. Holder's Tortugas data. Finding these creatures once in a hundred years on a given Key seems about normal to me.

References

Babbitt, L. H. 1951. Courtship and mating of *Eumeces egregius*. *Copeia* 1951(1): 79.

Mount, R. H. 1968. *Eumeces egregius*. *Catalogue of American Amphibians and Reptiles* 73: 1–2.

GROUND SKINK
(Scincella lateralis)

These are very quietly moving, nervous lizards seen most frequently crawling about in leaves or on the ground, seldom climbing.

Hobart Smith, 1946

The official common name seems rather silly since all of our skinks are frequently or exclusively found on (or in) the ground. I am not too pleased with the scientific name *Scincella* either. My friend and colleague Dr. Allen E. Greer of the Australian Museum, Sydney, is the acknowledged authority on skink taxonomy. He has fragmented the old genus *Lygosoma* into an assortment of names, but the definitions seem vague and inconsistent to me. This is the very sort of situation I deplore in *Natrix* or *Anolis,* for example, but I have closely studied those groups in the field, museum, and literature. With them I feel secure in my views based on first-hand knowledge. I have not studied skinks and will let *Scincella* stand for the moment. I provide you references if you wish to pursue the matter.

These are little brown lizards, growing a bit over five inches (13 cm) total length. The maximum head-body length is 5.9 cm. The sexes are quite similar. The middle of the back is chocolate to bronze brown, often shading to copper on the head. The sides are dark; one may think of this as a pattern of broad, near-black lateral stripes. The belly is often tinged with lemon yellow and the tail may be bluish. I have seen one from Big Torch Key that was heavily stippled with blackish middorsally, and another from Middle Torch that had a decidedly pinkish tail. This one might have been thought a red-tail skink but for its eyelids.

Ground skinks have windowed lower eyelids. This condition is intermediate between the opaque-scaled eyelids of *Eumeces* skinks and the complete, immobile, transparent scale called the spectacle seen in snake-eyed skinks (and many geckos and snakes). The ground skink can see with its eyes closed.

Fitch and Greene studied ground skinks from Kansas to the Gulf Coast. They found females began laying clutches of one or two eggs when they had attained a snout-vent length of 36 mm to 41 mm. Clutch size increased to an average of four for females over 48 mm, snout-vent, but declined again for the largest sizes, up to 59 mm. Six eggs was the largest clutch they found.

Each female laid three clutches a season, as a rule, spaced at about five-week intervals, from April through August. The eggs took about one month to hatch.

Brook studied the species on the Florida mainland and found they lay one to five (usually three) eggs with thin but tough shells in humus or rotten wood. Laying time was June to early August, hatching time from late August through September.

None of those data given above are necessarily applicable to the Keys, and I have no data specific for our islands. Ground skink mothers apparently do not guard their nests.

Carr (1940) provides strong circumstantial evidence for autophagy in these skinks. They actually eat their own tails. Ballinger (1981) discusses adaptive values autonomy—the tail breaking off—and fat content of lizard tails. Do skinks use their tails as a food resource in hard times? What a fascinating notion.

Smith (1946) notes: "A study of the variation very probably will reveal a number of geographical races." No one seems to have done that study yet.

I have examined museum specimens from Key West, South Saddlebunch, Cudjoe, Summerland, Water Key north of Big Torch, Carolyn Key west of Middle Torch, Big Pine, and Indian Key. Carr (1940) reports on three FSM specimens from the Marquesas. Duellman and Schwartz add Stock Island, Boca Chica, Sugarloaf, Summerland, Long Key, Lignumvitae, and Upper and Lower Matecumbe.

I have examined live specimens on both Big and Middle Torch Keys. We encounter one about every seven times we turn out 100 m² of cover in hammock L9/16 on Middle Torch. I have found these skinks locally far more common than that, especially in buttonwood transition and low hammock.

References

Ballinger, R. E. 1981. Can predator defense be tributive or toxins nontoxic? *American Naturalist* 117: 794–95.

Brook, G. R. 1963. Food habits of the ground skink. *Quarterly Journal of Florida Academy of Sciences* 26: 361–67.

————. 1967. Population ecology of the ground skink, *Lygosoma laterale. Ecological Monographs* 37: 71–87.

Fitch, H. S., and H. W. Greene. 1965. Breeding cycle in the ground skink, *Lygosoma laterale. University of Kansas Publications, Museum of Natural History* 15(11): 565–75.

Greer, A. E. 1974. The generic relationships of the scincid lizard genus *Leiolopisma* and its relatives. *Australian Journal of Zoology*, Supplement, Series 31: 1–67.

————. 1977. The systematics and evolutionary relationships of the scincid lizard genus *Lygosoma. Journal of Natural History* 1977(1): 515–40.

RACERUNNER
(AMEIVA SEXLINEATA)

*The racerunner depends primarily on speed to escape from its
enemies. . . . Ordinarily they detected a human intruder before he
was aware of the lizard's presence; the lizard might be noted merely
as a flash of motion as it darted away to shelter.*

HENRY S. FITCH, 1958

Patterned in grey, brown, and cream—often with a bluish tail, especially
when young—this lizard has three bold light stripes on each side of the
back: six stripes total. The belly is blue or bluish-grey. The dorsal scales
are tiny granules, imparting a sandpaper look—definitely not shiny. The
belly is covered by large plates.

These are true sun worshippers. They dwell almost exclusively on the
ground and do not even exit their burrows on cloudy days. They bask in
the sun until they achieve their optimum temperature of about 40°C or
104°F. Then they take off. They are fast, nervous, constantly active
hunters of insects and crustaceans. They are most common on the broad
expanses of open oölite in transition zones, or on sand strands and berms
with little vegetation. They belong to a group within the family Teiidae, of
which they are our only representatives, extremely widespread and abundant in the deserts of the southwest and the arid lowlands of the Antilles.
There are great radiations of species and subspecies.

Standard books call this species *"Cnemidophorus sexlineatus."* The
anatomist and paleontologist Dr. Richard Estes, working on Florida fossils
in 1963, pointed out that no hard anatomical features exist to justify the
generic name *"Cnemidophorus."* The putative difference is the retrac-
tability of the tongue. In *Ameiva, sensu stricto,* the tongue is more or less
retractile; in *"Cnemidophorus"* it is not. However, the actual difference, a
matter of the extent of integumentary sheathing on the tongue base, is
variable and equivocal. I had concluded that *"Cnemidophorus"* was invalid
on soft anatomical grounds quite independently of Estes, while studying
species from the Lesser Antilles and Swan Islands in the 1960s. The
"Cnemidophorus" condition pops up in various otherwise unrelated species of *Ameiva.* The German biologist Wagler coined *"Cnemidophorus"* in
1830. The French Baron Georges Cuvier penned *Ameiva* in 1817. By the
ironclad law of priority, *Ameiva* wins and stands as the correct name.

Our species seems to have been the subject of intensive study by

Stanley Elwood Trauth while he was at Auburn University in about 1980. Trauth has not published his results where I can find them, except for reproductive data.

Trauth studied 195 egg clutches from ten states—but not one from Florida. Dr. Henry Fitch studied this species in remote Kansas. Trauth found clutches varied from two to four eggs; Fitch recorded a high of five. Three is average. Females may lay two or three clutches per season in a hole in sand, loam, humus, or rotting wood. The eggs are soft-shelled and white. Laying to our north begins in late May and proceeds through August. Trauth found drought reduced clutch size. These reproductive data are probably not wholly valid for the Florida Keys populations.

The abstract—all I have seen—of Trauth's doctoral dissertation proposes recognition of some subspecies including one from Florida. By implication this one would not include the Keys populations, but would be centered on the midpeninsular area: the old interglacial Highlands island discussed under The Land, above.

Racerunners attain nine-and-a-half inches (24 cm) total length; most of that is tail. The head-body maximum is 7.6 cm. Males get larger than females, but the sexes are quite similar. It would be fascinating to study water balance in these hot, dry creatures: their behavior and habitat suggest almost no chance to drink fresh water. They do not even emerge until after the dew has burned off.

Racerunners are locally common in the Keys, but more easily seen than caught. I have seen museum specimens from Key West and Summerland. Carr (1940) reports an FSM specimen from Boca Grande. I have color and habitat notes on specimens observed alive on South Saddlebunch, Cudjoe, Little Torch, Big Pine, No Name, and Little Pine. Duellman and Schwartz report Stock Island, Boca Chica, Sugarloaf, the Newfound Harbor Keys south of Big Pine, Upper and Lower Matecumbe, and Key Largo. In all my months of fieldwork on Middle Torch Key I have yet to see this species there.

References

Estes, R. 1963. Early Miocene salamanders and lizards from Florida. *Quarterly Journal of the Florida Academy of Sciences* 26(8): 234–56.

————. 1969. Relationships of two Cretaceous lizards (Sauria, Teiidae). *Breviora* 317: 1–8.

Fitch, H. S. 1958. Natural history of the six-lined racerunner. *University of Kansas Publications, Museum of Natural History* 11(2): 11–62.

Trauth, S. E. 1980. Geographic variation and systematics of the lizard. *Cnemidophorus sexlineatus* (Linnaeus). *Dissertation Abstracts International* 41(1): 8013887.

————. 1983. Nesting habitat and reproductive characteristics of the lizard *Cnemidophorus sexlineatus* (Lacertilia: Teiidae). *American Midland Naturalist* 109(2): 289–99.

KEY MUD TURTLE
(KINOSTERNON BAURII BAURII)

The form here described is placed on record in the literature under the name of the eminent osteologist, Dr. George Baur of Clark University.

SAMUEL GARMAN, 1891

Garman described this species, popularly known as the striped mud turtle—a name fitting the mainland subspecies—from Key West and Cuba. No one nowadays believes the Cuban specimen represented a natural population; it was probably a pet, signaling the regular traffic between Key West and Havana, across a mere 90 miles, we once enjoyed.

These are small turtles, up to almost five inches (12.2 cm) shell length. They live in fresh to brackish ponds in the hammocks and will leave them if salt begins to concentrate in them as they dry up in winter. They are quite different physiologically from their mainland cousins, and have been well studied by Dr. William Dunson of Penn State University. Their taxonomy has been controversial. Uzzell and Schwartz (1955) were the first to perceive that the Lower Keys animals, nominate *baurii*, were consistently different from mainlanders and that the boldly striped form, *palmarum*, first described from Everglades, was found throughout the mainland range of the species.

Iverson (1978) studied aspects of pigmentation and morphometrics of carapace and plastron. He concluded that, despite striking differences between northern and southern populations, "subspecific recognition of a Lower Keys population . . . is unjustified." He based his opinion on the fact that shell proportions vary clinally, from north to south, and that some northern individuals (from the Gulf Hammock, Georgia, and South Carolina) are as dark as specimens from the Lower Keys.

I have examined most of the specimens seen by Iverson, including all the University of Florida material (with special attention to Gulf Hammock specimens) and all the individuals regarded by Iverson (*in litt.*) as compromising the subspecies. In addition, I have examined 18 more *palmarum*—two in the Florida State University collection from Lake and Hendry counties, Florida, and 15 in the University of Georgia collection (some formerly of Georgia State University) from Alachua, Dade, Lake, Monroe, and Hernando counties, Florida; Barnwell County, South Carolina; and MacIntosh and Wilcox counties, Georgia. I have examined, measured, photographed, and released two additional Middle Keys individuals (both females) from Key Vaca.

I have similarly examined and measured (but often not photographed) about 100 live individuals from the Lower Keys. I have accumulated eight more Lower Keys specimens salvaged as road-kills or that died in captivity. All are now in the MCZ. I am especially indebted to Dr. William Dunson for some of the latter specimens and many of the live ones, as well as advice, and many habitat data. I am additionally indebted to Dr. John Iverson, who provided a copy of his computer printout on shell morphometrics and advice on individual specimens to examine.

I follow Simpson (1961, see citation under The Fauna, above) in recognizing as valid subspecies those geographic variants that can be readily identified at a high level (at least 75 percent) without recourse to locality data. Like Simpson, I believe that steps in clines of geographic variation make fine places to delimit subspecies ranges. The cline in plastron width is sharply stepped between the Everglades and the Lower Keys. I found that every specimen of *K. baurii* I examined could be assigned correctly to subspecies without recourse to locality data, and over 80 percent could be correctly assigned on the basis of ostentatious color characteristics alone.

Here are revised diagnoses for the forms:

K. b. baurii: carapace dark with three dorsal light lines vaguely indicated or almost obliterated by dark pigment; mandible dark and without bold yellow streaks or blotches; plastron narrow; measured at the abdominal-femoral suture, 34–38 percent of carapace length in males and 37–43 percent in females.

K. b. palmarum: carapace variable, usually boldly striped; mandible variable, usually boldly streaked or blotched with yellow; dark individuals have wide plastrons: measured at the abdominal-femoral suture, 37–42 percent of carapace length in males and 43–51 percent in females.

While it is possible for a dark *palmarum* male to have a plastron as narrow as 37 percent of carapace length, and thus identify as nominate *baurii*, I can find no such specimen. Among females examined there was no overlap. The female nominate *baurii* with the widest plastron had a width 42.8 percent of carapace length; the only female *palmarum* with a narrow plastron from a population containing dark individuals (southern Georgia) had a width 43.4 percent of carapace length. Thus, 100 percent of *K. baurii* I examined could be correctly assigned to subspecies.

K. baurii from the Upper and Middle Keys identify as *palmarum* on color characters; their plastrons are narrow and widely overlap the proportions of nominate *baurii*. They may be thought of as intergrades, or as *palmarum* at the extreme end of its cline in proportions.

Iverson (1978) suggested that the dark coloration of *K. b. baurii* might be staining from acid waters. However, pond waters on the oolite limestone Lower Keys are characteristically circumneutral. Thirteen ponds known to harbor mud turtles were checked between Little Pine Key and Stock Island. They varied in pH from 6.5 to 7.9 (average 7.2); rainfall on 28 January 1979 was pH 5.0 at Little Torch Key.

Several *K. b. baurii* have been maintained in captivity in dechlorinated tap water varying from pH 5.5 to 7.5 for periods of several weeks to three years. Their coloration and pattern was wholly unaffected. I conclude that the dark pigmentation of nominate *baurii* is genetically controlled, not environmentally induced. Whether this dark coloration is adaptively valuable, I cannot presently determine. Certainly many of the most obvious differences between many geographic races of animals are adaptive. This in no way detracts from their value as taxonomic characters.

Kinosternon baurii baurii has apparently been extirpated from its type-locality, Key West. Dunson (1981) found the total population on southern Summerland to be 219–274 individuals, while pristine Johnston Key supported fewer, only 41–52. They could not tolerate salinities higher than 15 ppt, less than half that of seawater. Dunson pointed out ". . . a large proportion of extant *K. b. baurii* are on private property where they could be exterminated in the near future by real estate development."

A few of these turtles occur on lands within National Key Deer Refuge or Great White Heron National Wildlife Refuge, but no large populations in optimal habitat are so protected. In fact, severe hurricane damage or such management techniques as control burning (if done when the fresh marshes are dry) could possibly extirpate these peripheral and marginal populations.

The optimal habitats of *K. b. baurii* are small fresh or slightly brackish ponds in or at the edges of hardwood hammocks, where the land is highest and the fresh water lens is best developed. Effective mangrove and coastal wetlands protective legislation has forced most of the real estate development in the Lower Keys into these hardwood hammocks. At the present rate of destruction, the optimal habitats of *K. b. baurii* will predictably be gone in about 15 years; the form could easily be extinct by the year 2000.

The taxonomic controversy over these mud turtles is extremely instructive. Iverson believed the subspecies invalid because variation in shell proportions was clinal and, on principle, one does not name two ends of a cline. I value the striking, if modal, difference in color, which runs counter to the shell proportion cline and could be argued to indicate character divergence: the most geographically proximate populations are the most distinct. Such character divergence argues for full species status.

At the other end of the range of *K. baurii,* in Georgia and South Carolina, Trip Lamb has studied the morphology of the species where it meets and overlaps the closely related eastern mud turtle, *Kinosternon sub-*

rubrum. Here the two species are so similar that it requires sophisticated analysis of discriminant functions to tell them apart. Iverson was fooled by many specimens and misidentified them at species level. Lamb's papers are excellent because they provide the precise discriminant functions, involving six measurements (five for each sex) that work. It is much harder to identify these northern animals to species than it is to identify any *K. baurii* to correct subspecies.

Why then do we claim *baurii* and *subrubrum* are full species, while the much more readily distinguished nominate *baurii* and *palmarum* are but subspecies? Because of their apparent biology.*K. b. palmarum* and *K. subrubrum* occur together and do not seem to interbreed. With care and diligence, you can separate every known specimen.

K. b. baurii and *palmarum* nowhere occur together. The few known Upper and Middle Keys populations are intergradient. They contain individuals that usually identify as *palmarum* on color characters, but nominate *baurii* on shell proportions. Some Middle Keys specimens identify as nominate *baurii*. These turtles are evidence that speciation is as yet incomplete between mainland and Lower Keys mud turtles. If the distinctions were absolute—no overlap even potentially in shell proportions of dark individuals of *palmarum*—and if the Upper and Middle Keys intergrades did not exist, then *baurii* and *palmarum* would qualify as full species. It is ironic but true that, in this case, subspecies are more different from each other than are full species.

These turtles are listed as state-threatened by Weaver (in McDiarmid, 1978). He gives a diet for mainlanders. Ours are largely carnivorous scavengers or insectivorous. They lay up to four eggs in a hole excavated on land, wherever the female can find soil enough to dig it. She may lay three clutches a season, beginning in April. I have examined museum specimens from Key West, Stock Island, Saddlebunch L10/10, Cudjoe, Summerland, Johnston, all three of the Torch Keys, and Big Pine.

References

Dunson, W. A. 1979. Salinity tolerance and osmoregulation of the Key mud turtle, *Kinosternon b. baurii*. *Copeia* 1979(2): 548–52.
————. 1981. Behavioral osmoregulation in the Key mud turtle, *Kinosternon b. baurii*. *Journal of Herpetology* 15(2): 163–73.
Garman, S. 1891. On a tortoise found in Florida and Cuba, *Cinosternum Baurii*. *Bulletin of the Essex Institute* 23(9): 1–3.
Iverson, J. B. 1978. Variation in striped mud turtles, *Kinosternon baurii* (Reptilia, Testudines, Kinosternidae). *Journal of Herpetology* 12(2): 136–42.

Lamb, T. 1983. On the problematic identification of *Kinosternon* (Testudines: Kinosternidae) in Georgia, with new state localities for *Kinosternon bauri*. *Georgia Journal of Science* 41: 115–20.

_____. 1983. The striped mud turtle (*Kinosternon bauri*) in South Carolina, a confirmation through multivariate character analysis. *Herpetologica* 39(4): 383–90.

Uzzell, T. M., and A. Schwartz. 1955. The status of the turtle *Kinosternon bauri palmarum* Stejneger with notes on variation in the species. *Journal of the Elisha Mitchell Scientific Society* 71(1): 28–35.

FLORIDA BOX TURTLE

(TERRAPENE CAROLINA BAURI)

The oldest known fossils of the genus are of Pliocene Age. They are fully differentiated as to both generic characters and species group characters, and thus give no clues to the origin either of the genus or the species groups.

W. W. MILSTEAD, 1969

In his fascinating and meticulously detailed account of the box turtles, Milstead (1969) paints a picture of constant movement of species and subspecies over time. Since the Pliocene ended a million years ago, and the Keys only became connected to mainland Florida about fifty thousand years ago, our box turtles are a perfect example of such movement. Box turtles are apparently quite inept at over-water crossings and, I believe, must be stranded on the Keys since they walked here during the Wurm glacial maximum.

I have problems with Milstead's taxonomic assessments, which seem to me to be extreme hair-splitting. As is usual with geographically contiguous, intergrading subspecies (the best kind), all characters do not always break sharply along exactly the same lines. The major diagnostic characters break sharply, but other features blend more gradually. Thus one can see a slight trend toward the shell proportions of Gulf Coast *T. c. major* in our Key turtles, and even a hint of influence from the extinct *T. c. putnami* in the ratio of the plastral lobes. None of this impairs their basic identity as *T. c. bauri* in my view.

Florida box turtles are mahogany brown to nearly black with rich yellow lines or spots radiating outward from the original laminus of each big carapace scute. They have yellow head and neck stripes and the plastron is bold yellow and brown. They may attain a shell length of nearly seven inches (17.5 cm). Males have more elongate shells on average than females, with a more flared back edge, and a concave plastron (bottom shell). Hatchlings and young have a broad yellow vertebral stripe. The number of toes is variable on the hind foot; there may be three or four.

Females excavate a nest with their hind feet (like all turtles) for egg deposition sometime from April through June (Carr, 1952). From one to nine eggs—usually five—are laid; they take about ninety days to hatch. As is the rule with turtles, there is no parental care; females never look back on their nests. Density of adults may exceed two per hectare.

Box turtles have suffered grievously from development. They are hopeless homebodies with special microhabitats within their home ranges where they seek shelter and moisture. Automobiles on the roads that now dissect the hammocks and pinelands are their doom. Transporting box turtles away from their home ranges usually kills them too: they will try to get home again, and are usually run over in the process. They do fairly well in captivity if fed a mixed diet of fruit, vegetables, and meat. In nature they are highly insectivorous, especially when young.

The only museum specimens I have examined are three from Howe Key, one from Big Pine, and an empty shell salvaged on Middle Torch. Many people pick up box turtle shells found in the woods and keep them as souvenirs. I have examined such from Lignumvitae, Howe Key, and Big Knock-'em-Down. I have seen, but not molested, live individuals on Howe, Big Pine, Middle Torch, Cudjoe, and Key West. Duellman and Schwartz add Summerland and Key Largo.

Box turtles are utterly American with no close relatives anywhere else. They are both pretty and charming. Many people cannot resist taking them from their natural haunts to be kept in captivity—which renders them just as genetically dead as if you pickled them in formaldehyde and put them in a museum. When I was a child there were fewer roads, and many, many fewer people. I kept pet box turtles. It seemed all right then. I released mine back where I had caught them, but after a year or so in captivity they had probably been replaced and could no longer fit into their original homes. In 1950 that hardly mattered, there were so many of them.

Today they are in bad shape—a mere remnant of their former numbers. We must sacrifice many things to have our huge and burgeoning human population—things I would much rather have back again. Box turtles common enough to be conscionably kept as pets are among those things. Please leave them where you find them so that their species may have some chance against the rising tide of our own. If you salvage an empty shell, send it to a museum.

Reference

Milstead, W. W. 1969. Studies on the evolution of box turtles. *Bulletin of the Florida State Museum* 14(1): 1–113.

MANGROVE TERRAPIN
(MALACLEMYS TERRAPIN RHIZOPHORARUM)

In June, 1904, Henry W. Fowler, the distinguished naturalist from the Philadelphia Academy of Natural Sciences, together with a colleague, cruised through the Florida Keys to search for tree snails of the genus Liguus. *Incidental to their snail searching, they discovered a smallish, unobtrusive turtle on Boca Grande.*

ROGER WOOD, 1981

Two very different sorts of mangrove terrapins appear to live in the Keys; their relationships, ecology, and population biology are presently being investigated by Dr. Roger Wood of Stockton State College, Pomona, New Jersey. While we await definitive published results, I can summarize what I know from field experience, a couple of Wood's typescripts, and a typescript on fieldwork by David Ryan (1981).

Diamondback terrapins occur in the mangrove habitats of Everglades, the Florida Bay Keys (Wood lists Bottle, Tern, Park, and Manatee Keys), and Key Largo (*fide* Duellman and Schwartz). I have seen dozens of museum specimens of this form, mostly salvaged from eagle nests by Dr. William Robertson and deposited at FSM, University of Florida. There are also pickled specimens at MCZ, Harvard, and no doubt elsewhere. I agree with Carr (1952) that these Florida Bay turtles are basically intermediate between the peninsular east coast *Malaclemys terrapin tequesta* and *M. t. macrospilota* of the Gulf coast. Wood does not agree that these are the same as Henry Fowler's mangrove terrapin, *M. t. rhizophorarum*, and neither do I.

A huge gap in the range of terrapins involves the Upper Keys below Key Largo, the Middle Keys, and most of the Lower Keys. Fowler's *rhizophorarum* came from the sand Keys west of Key West, in what is today Key West National Wildlife Refuge. Roger Wood obtained the necessary federal and state permits to work on turtles in these Keys, and set out in 1981 with Bill Ford as his guide, and a team of Earthwatch volunteers, to find the animals Henry Fowler had described. They managed quite handily, although the animals are genuinely rare and undoubtedly endangered unless Refuge regulations and state laws are zealously enforced. This subspecies is explicitly protected, and listed as rare by Godley (in McDiarmid, 1978). Godley's account confounds the two sorts Wood recognized and does not acknowledge their geographic disjunction.

The true *rhizophorarum* is known from the Marquesas, Boca Grande, Barracouta, and Man, Archer, and Cottrell Keys on the basis of museum specimens (MCZ, FSM, and NMNH). Sight records for as far east as the Content Keys exist, but efforts to locate populations east of Key West, or in the Tortugas, have proved futile so far.

This turtle is somber, dark, and shows little pattern except for nearly black pigment along the seams of the otherwise yellowish plastral scutes. It lacks the head and neck spots and stripes, and the "striped pants," characteristic of Florida Bay individuals. These features are cited in all the field guides and popular literature because no one before Roger Wood really went back, located, and accurately described the real *rhizophorarum*. Diagnosis of the subspecies will depend on shell and scute proportions, not coloration.

Geographic variation in terrapins is considerable, as befits a species occurring from Texas and the outer Keys north to Massachusetts. However, I find individual variation swamps most of the field guide characters given in books. Furthermore, some of the nominal subspecies, such as "*littoralis*," are probably intergrades with the fresh water sawbacks and map turtles popularly put in a separate (indefensible) genus called "*Graptemys*." The entire taxonomic arrangement fairly screams for intelligent revision (see Lazell, 1976, for discussion).

Mark-recapture studies by Wood and his assistants indicate a total population of real *rhizophorarum* of only a few hundred. The biggest individuals are said to be about nine inches (less than 23 cm) shell length, but that is for the species as a whole. Females get much larger than males and have broad heads. Small-headed males have large tails when mature. In the North terrapins can live for decades.

Aspects of the biology of *rhizophorarum* are puzzling and unresolved. Most terrapins lay eight or ten (up to 16) eggs in a clutch in sand well above tide line. They may lay two or three clutches per year in temperate climes, beginning in late spring. The eggs take about 90 days to hatch. We do not know, and in many cases cannot imagine, where mangrove terrapins lay their eggs: the little berms on most of their islands are too low to avoid flood tide drown-outs. The Marquesas are an exception, but use of this nesting habitat implies migration for most of the other known individuals. The sex ratio Wood found was four to twenty females per male. Most of Robertson's eagle nest specimens from Florida Bay were males. The sex ratio among Massachusetts adults is one to one. We know that nest temperatures affect sex ratio of hatchlings, and that warmer nests produce females. Explaining the data will require some wonderful hypotheses and arduous field work (now under way by Peter Auger of Tufts University).

Water balance was studied in Florida Bay terrapins by Dr. William Dunson (1970). True *rhizophorarum* are apt to live in far more saline waters than more northern individuals seem able to. Wood found *rhizophorarum* preferred water 45–70 ppt (seawater is about 36 ppt). It is difficult to imagine hatchlings surviving in such salty conditions.

Wood did find these turtles active in heavy rain. Perhaps, like mangrove snakes (which see), they are able to utilize rain water for their physiological needs and thus exploit food resources in hypersaline swamps. Wood found they ate primarily small aquatic snails.

I did not participate in the early work done by Wood, Ryan, and others in 1981. However, I went out to the sand keys with Bill and Fran Ford in January 1985. In no time we found (and left right there) two of Wood's marked turtles from four years earlier: they had scarcely moved. The radio transmitter study done by Wood and Ryan in 1981 indicated home ranges of about 750 m², which is much smaller than the home range of northern terrapins (for a striking contrast, see home range of rice rats, above).

The diamondback terrapin was once nearly extirpated, at least in the Chesapeake region, by the gourmet restaurant trade. The meat is as good as chicken, but much more difficult to prepare acceptably. Even if our mangrove terrapins were not rigidly protected by federal and state laws, no arguable case could be made for exploiting them as a food resource, although consumption by humans may have generated the apparent gap in the range.

References

Dunson, W. A. 1970. Some aspects of electrolyte and water balance in three Florida estuarine reptiles: the diamondback terrapin, American crocodile, and "salt water" crocodile. *Comparative Biochemistry and Physiology* 32: 161–74.

Fowler, H. W. 1906. Some cold-blooded vertebrates of the Florida Keys. *Proceedings of the Academy of Natural Sciences of Philadelphia* 58(1): 77–113.

Lazell, J. 1976. *This Broken Archipelago*. New York: Quadrangle, The New York Times Book Co.

Ryan, D. 1981. The mangrove diamondback terrapin, *Malaclemys terrapin rhizophorarum* in the Florida Keys. Unpublished typescript, 20 pp. Available from The Conservation Agency, Jamestown, Rhode Island.

Wood, R. C. 1981. Search for the mangrove terrapin. Typescript report to *Earthwatch*, Watertown, Massachusetts.

————. 1981. The mysterious mangrove terrapin. *Florida Naturalist* 54(3): 6–7.

INTRODUCED TURTLES

Due to the popularity of turtles as pets, non-native species turn up occasionally. . . . Few of these result in breeding populations since most releases are juveniles and few in number.

ASHTON AND ASHTON, 1985

Over the years I have observed a number of terrestrial and fresh water turtles in the Keys that I have assumed were introduced exotics. Occasionally, I have salvaged dead specimens, but I have not purposefully collected any. My opinion of their true status is nothing more than a gut feeling; so far as I know, no one has ever given serious consideration to any of these as natives here. None of those listed below were noted by Ashton and Ashton (1985) as occurring even in Monroe County. They did not consider the MCZ collection, where most of my specimens are deposited, and missed at least one specimen in FSM, University of Florida.

This list can scarcely be complete. I have only included species I have actually seen. I am informed that the old record for *Kinosternon subrubrum* (a species of mainland mud turtle), for example, was based on a misidentification of *K. b. bauri,* not on an actual specimen.

Gopher tortoise (*Gopherus polyphemus*). These ponderous creatures reach a shell length of over 14 inches (nearly 37 cm). They are somber in color and live in long burrows that they excavate. A population lives in Monroe County, at Cape Sable in Everglades National Park. Pet gophers turn up most anywhere. I have seen them in Key West. A few were maintained on Little Torch Key some years ago, and one or two escapees may still live there. This tortoise is listed as a threatened species (Auffenberg in McDiarmid, 1978) and is protected by law. The direct threats are butchering gophers for their meat, poisoning them with gasoline during rattlesnake roundups, and highway mortality. Indirect, but probably even more dire, threats are habitat destruction through agricultural and urban development and capture for the illegal pet trade. Excellent accounts of this species are given by Ashton and Ashton (1985) as well as Auffenberg's article (cited above) and the references therein.

Peninsula cooter (*Chrysemys floridana peninsularis*). This is a large pond and river turtle, common and native in the Monroe County Everglades. It may attain a shell length of nearly 16 inches (40 cm). It is frequently eaten by people, and may be transported for this purpose. Cooters make poor pets because of their aquatic habits and need for just the right exposure to sunlight for calcium metabolism. They are basically yellow and blackish. The carapace of old adults can become very dark, the narrow yellowish lines virtually obsolete. The plastron is nearly white. Yellow stripes on the head form "hairpin" markings on the top of the back of the head. I have seen this species on Big Pine and salvaged a large, long-dead female containing shelled eggs on Stock Island in June of 1977. I put her in FSM. Because of the decomposed state of this specimen and the difficulties involved in identifying cooters in general, I am not positive of my identification of this individual. I doubt she represented a breeding population. Superb, detailed accounts of this and the following species are given by Carr (1952).

Florida redbelly turtle (*Chrysemys nelsoni*). This is another large pond and marsh turtle, but it frequents brackish water. It is a common native of the Everglades and is the candidate I regard as most likely to be native in the Keys—at least on Key Largo. It does not grow quite as large as the cooter: to just over 13 inches (34 cm). Young specimens are brightly patterned with reddish markings on their brown to blackish shells. The plastron is especially reddish, shading to orange or rosy-red at the edges. Yellow stripes are on the near-black skin of the head and neck; these do not form the hairpin markings seen on the cooter, but instead make an arrow-like marking, beginning between the eyes, at the snout. I salvaged the empty shell, with scutes, of a big female (27.5 cm) on Big Pine Key in January 1981. Living individuals occur on Key Largo. Pond turtles of one sort or another are regularly visible at the Blue Hole on Big Pine, within National Key Deer Wildlife Refuge.

Florida softshell turtle (*Trionyx ferox*). This large, leathery, snorkel-nosed species has been introduced into the Big Pine Blue Hole. They like to haul out and bask in the sun and may be closely approached. They have long necks and can bite ferociously: that is what *ferox* means. I do not know if they can find nesting habitat on Big Pine and thus reproduce.

Some effort to identify our odd turtles would resolve questions of who is where; the possibility of reproducing—even possibly native—populations should not be discounted out of hand.

Reference

Wilson, L. D., and L. Porras. 1983. The ecological impact of man on the south Florida herpetofauna. *University of Kansas Special Publication* 9: 1–89.

GREEN TURTLE

(CHELONIA MYDAS MYDAS)

*While once probably abundant in Florida, the current population is
estimated to contain no more than 50 mature females.*

FRANK LUND, 1978 (IN McDIARMID)

*The green turtle is a staple in Key West, where large numbers are
brought in for shipment and for the local market.*

ARCHIE CARR, 1940

This is the species most people think of when they refer to a sea turtle. It
has been exploited for centuries for eggs, meat, and calipee. It is the
calipee, a fatty substance clinging to the inside of the shell, which gives
these turtles their common name. Green turtles are not green—they are
horn or tortoise-shell colored. The calipee is rendered for soup, and the
soup is greenish: green turtle soup. It is the codification of a misunder-
standing to call the turtle green.

The Key West turtle kraals closed down as a commercial slaughter-
house in 1972 with the advent of endangered species awareness and laws.
CDR William R. Ford, U.S.N. (Ret.) had managed the operation since
1965, at which time it was owned by the A. Granday Canning Company,
one of Thompson Enterprises, Inc. In 1967 Granday sold out to Sea Farms,
Inc. Thousands of green turtles were processed in Key West between 1965
and 1972, but Bill Ford does not believe the operation was a mainstay of
the Keys economy. The turtles were caught by Caymanian turtlers—from
Grand Cayman Island north of Jamaica—on the banks off Nicaragua, or
rarely Mexico. The few Conchs employed by the outfit were also employed
in shrimp cleaning and packing, so the Endangered Species Act did not
cost them their jobs. Local people still caught (and no doubt continue to
catch) a few green turtles for their own consumption, but local turtles
were never a part of the larger commercial operation.

Green turtles are not rare in Keys waters. Holder, in his 1892 book cited
under Reptiles and Amphibians above, knew this species well at the
Tortugas, where it was frequently held in captivity and slaughtered for the
meat. Although Lund (in McDiarmid, 1978) does not acknowledge it,
green turtles nest regularly in the Keys at the same beaches used by
loggerheads (which see, below), albeit in much smaller numbers. I have
seen the shell of a big female killed while she nested on Lower Sugarloaf.

Many Key West specimens exist—mostly skulls—but these are probably all from butchered turtles taken off Nicaragua. Salvaged specimens, now in major museums, which I have examined are young individuals from Lower Sugarloaf and Boca Grande.

Sea turtles are hard to identify until you get used to them. I find the presence or absence of a small, anterior costal plate, the shoulder scute, to be the best feature to look for first. Green turtles lack it: all the costal—or side—scutes (or plates) are large, as in most turtles. Hawksbills also lack this scute, but the other hardshelled sea turtles possess it. Hawksbills have *four* plates on the top of the head, on the snout and between the eyes, while green turtles have but *two*. Greens and hawksbills do occasionally hybridize, so a genuinely confusing individual may turn up.

Green turtles regularly attain shell lengths up to 60 inches (153 cm) and weights up to 280 lbs (127 kg). Conant (1975) says they may exceed 650 lbs (295 kg). An adult female lays about 100 eggs per clutch, and may lay as many as three clutches in a nesting season. In the Gulf and Caribbean population nesting takes place in spring and summer. Other stocks may be very different. Females usually nest every other year, or even every third or fourth year. At least that is what we customarily believe. Considerable evidence indicates that many females only nest once in a lifetime, and that those who return to nest in another season are lucky exceptions to the rule (Hughes, 1982).

Hatchling green turtles are quite carnivorous, but the adults browse on turtle grass, *Thalassia*. Like mammalian herbivores such as cattle, rabbits, and kangaroos (but quite unlike humans), green turtles have a modified digestive system in which they maintain a bacterial culture that breaks down cellulose.

A true gut fermentor, like the green turtle or the cow, depends on the bacteria to render cellulose as ethyl alcohol and fatty acids, which supply the animal's caloric (energy) needs. Excess bacteria are digested also, provide protein, and thus essential amino acids wanting in the plant material. Green turtles, when adults, still like to eat some invertebrates, like crabs and jellyfishes, and will take fish in captivity. They ingest a lot of animal protein when eating turtle grass, too, because small organisms encrust the grass blades. They are not strict vegetarians.

The life history of sea turtles contains a big blank, popularly called "the lost year." We do not certainly know what happens to them after they swim out to sea from the beaches where they hatched. After far more than a single year in most cases, they reappear as subadults, often in places remote from the grazing grounds of their parents. Some baby turtles spend at least part of their "lost" year(s) in floating rafts of *Sargassum,* an abundant warm water weed.

Sea turtle biology has begun to receive the attention it has so long deserved in view of the economic importance of the animals in many human cultures. There has been a relative explosion of literature. For example, number three of the 1985 volume of *Copeia,* the journal of the American Society of Ichthyologists and Herpetologists, contains a baker's dozen of sea turtle articles. The ponderous 583-page tome edited by Dr. Karen Bjorndal, cited below, contains more than sixty papers by noted sea turtle authorities. I shall refer to some of these under the other species, below.

References

Bjorndal, K., ed. 1982. *Biology and Conservation of Sea Turtles.* Washington, D.C.: Smithsonian Institution Press.

Carr, A. 1967. *So Excellent a Fishe.* Garden City, N.Y.: Natural History Press.

_____. 1982. Notes on the behavioral ecology of sea turtles. In *Biology and Conservation of Sea Turtles,* K. Bjorndal, ed., pp. 19–26.

Hughes, G. R. 1982. Nesting cycles in sea turtles—typical or atypical? In *Biology and Conservation of Sea Turtles,* K. Bjorndal, ed., pp. 81–89.

Rebel, T. P. 1974. *Sea Turtles and the Turtle Industry of the West Indies, Florida and the Gulf of Mexico.* Coral Gables, Fla.: University of Miami Press.

Thayer, G., K. Bjorndal, J. Ogden, S. Williams, and J. Zieman. 1984. Role of larger herbivores in seagrass communities. *Estuaries* 7(4A): 351–76.

Zieman, J., R. Iverson, and J. Ogden. 1984. Herbivory effects on *Thalassia testudinum* leaf growth and nitrogen content. *Marine Ecology Progress,* Series 15: 151–58.

HAWKSBILL
(CHELONIA IMBRICATA IMBRICATA)

Source of the tortoise shell of commerce, and still in demand as a luxury item even though plastics have replaced many of its former uses.

ROGER CONANT, 1975

This species is usually given the generic name *"Eretmochelys,"* but it hybridizes with *Chelonia mydas* both in captivity and in the wild, thus proving the two species are about as closely related as species can be. The hawksbill is handsomely patterned in brown and yellow and usually has large, overlapping, shingle-like scutes on the shell (imbricate means shingle-like).

Hawksbills do not get as large as greens: to a shell length of 36 inches (91 cm) and weight of 280 lbs (127 kg). Like the green, they lack a shoulder scute. However, they have *four* plates on the top of the head, on the snout and between the eyes—two pairs. Green turtles have only two scutes there—one pair.

The hawksbill is largely carnivorous throughout life and very much a denizen of coral reefs. I have encountered hawksbills of all ages and sizes on the same reefs, so I do not believe this species has a "lost" year or period after hatching when it occupies an unknown habitat.

Hughes (cited above) reports data indicating that hawksbills nest infrequently, every three to six years, if one nests a second time at all. I find no report of multiple clutches in a single season, and the egg count given is quite high for a single clutch: 160 eggs (Ashton and Ashton). Like other sea turtles, the eggs hatch in about two months.

Only two nests have been reported in Florida, from up on the east coast. However, hawksbills are not terribly rare in the Keys, and I believe some nest here. Charles Holder, back in 1892, reported adults present in the Dry Tortugas.

I saw a fine, big captive adult at the Key West turtle kraals in December 1985. The only museum specimens I have seen were subadults, except for skulls from "Key West" of suspect origin. The subadults are from Lower Sugarloaf, the south side of Coupon Bight on Big Pine, and Key West.

Reference

King, F. W. 1982. Historical review of the decline of the green turtle and hawksbill. In *Biology and Conservation of Sea Turtles*, K. Bjorndal, ed., pp. 183–88.

LOGGERHEAD
(Caretta caretta caretta)

This species is an omnivorous turtle which is often reported in temperate waters....

James Perran Ross, 1982

Even though severely depleted from its original abundance, this is the common sea turtle in Florida waters. Certainly hundreds (possibly over a thousand) of loggerheads nest each year in the Keys, although Lund (in McDiarmid, 1978) shows none at all. The principal nesting areas are the Dry Tortugas (where much basic work on the biology of these turtles was done at the Carnegie Institute laboratory), the Marquesas and other sand keys west of Key West, along Lower Sugarloaf, at Cocoplum Beach in Marathon, Long Key, and along Lower Matecumbe. I have also good records for southern Big Pine; a few loggerheads are apt to nest anywhere they can find digging sand to excavate a nest hole.

Loggerheads lay several clutches of about 100 eggs in the course of a nesting season, from April to August. A given female may nest six times in nine years, others may nest once in a lifetime. The eggs hatch in 60 to 100 days. Hooker (1911) found that hatchlings are repelled by the colors red, orange, and green, and attracted to blue. This should function to get them to water, except most hatch at night when colors are least visible. They also prefer dark to light—they are negatively phototropic. This would serve to propel them off the beaches and into the sea even at night.

Loggerheads are basically brown with reddish and yellow tones. They have a small shoulder scute as the first plate in the costal series on each side of the back. They may exceed 48 inches in shell length (122 cm) and weights of 500 lbs (227 kg). The southeastern United States is a stronghold for this species, which is heavily hunted over most of the rest of the world for meat and eggs. Because of good laws, enforcement, and the popularity of "turtle watches" on the nesting beaches, loggerheads now nest on our shores in far greater numbers than even two decades ago. They nest as far north as New Jersey and summer even in New England and Canadian waters.

There are hatchlings preserved in museums from "the Keys," with a specific locality for Upper Sugarloaf. There are other museum specimens from Long Key and Key West. Evidence suggests that hatchlings spend their "lost year" out in the *Sargassum* rafts of the mid-Atlantic.

References

Dodd, C. K. 1988. Synopsis of the biological data on the loggerhead sea turtle *Caretta caretta* (Linnaeus 1758). U.S. Fish and Wildlife Service Biological Report 88(14): viii + 110 pp. Does not include original research or much information on the Keys, but offers a comprehensive survey of the literature.

Hooker, D. 1911. Certain reactions to color in the young loggerhead turtle. *Papers from the Tortugas Laboratory* 3: 69–76.

Mast, S. O. 1911. Behavior of the loggerhead turtle in depositing its eggs. *Papers from the Tortugas Laboratory* 3: 63–67.

Richardson, J. I., and T. H. Richardson. 1982. An experimental population model for the loggerhead sea turtle (*Caretta caretta*). In *Biology and Conservation of Sea Turtles*, K. Bjorndal, ed., pp. 165–76.

Ross, J. P. 1982. Historical decline of loggerhead, ridley, and leatherback sea turtles. In *Biology and Conservation of Sea Turtles*, K. Bjorndal, ed., pp. 189–95.

KEMP'S RIDLEY
(LEPIDOCHELYS KEMPI)

Richard M. Kemp, of Florida, directed my attention to a peculiar Turtle, commonly called the "Bastard," found in the Gulf of Mexico, and said to be a cross between the Green and Loggerhead.

SAMUEL GARMAN, 1880

The type-locality of Kemp's ridley is Key West, where it is a rare species today. This ridley nested by the tens of thousands along the coast of the western Gulf from Texas through at least Tamaulipas as recently as 1947, but fewer than 5,000 adults are believed to survive today. This is the most critically endangered of all sea turtles.

Kemp's ridley resembles a loggerhead in possessing a shoulder scute, but is grey to olive in color and has a shell about as broad as it is long—very unusual in a turtle. This is a small species, not known to even reach 30 inches (74.9 cm) in shell length and a weight of 110 lbs (49.9 kg). Females lay an average of 140 eggs per clutch, with one or two clutches per season. There may be a pattern of nesting in consecutive years after two- to four-year intervals.

The life history of this species is hotly debated. The orthodox view, presented by Hildebrand, is that the entire life cycle is spent in the Gulf of Mexico. Thus, most authorities believe any ridley that goes around the tip of Florida has essentially committed suicide. As Hildebrand says: "Most of the turtles which pass through the straits are probably lost to the population."

I am the principal proponent of a different analysis of the available evidence. Recent work by Lutcavage and Musick in Virginia supports my view, developed in New England. I believe normal hatchlings of Kemp's ridley proceed as fast as they can swim—and the Stream can carry them—right through the Straits of Florida and out into the Atlantic. Here I believe they take up residence in floating *Sargassum* rafts. They are voracious carnivores all their lives, but the small prey in the *Sargassum* rafts soon fails to satisfy them. I believe they normally leave the *Sargassum* at a size of about five inches (13 cm) or more and swim westward to the continental littoral, where crabs and mollusks abound. To make this trip, they must cross the Gulf Stream. The smaller ones are carried farthest north by the Stream and fetch up in New England waters. Larger, older individuals—about 16 inches (40 cm) on average—reach the Virginia Capes. Near adult specimens reach the Georgia Bight, and full-sized adults reenter Florida waters, round the peninsula, swim through the Keys, and take up residence in the Gulf.

I am not sure what young ridleys do during the winter months. Do they migrate back out to the Sargasso Sea each fall? Do they hibernate even as far north as New England? We know ridleys hibernate in Florida waters and ordinary turtles hibernate in New England. Ridleys with sea lettuce (*Ulva*) growing on their shells have been dredged up in New England— good evidence they were at least attempting to hibernate.

At the present time no compelling evidence supports either the ortho- dox view or my heretical one. There is plenty of evidence: lots of ridleys can be found every summer in New England waters. It is the interpreta- tion of this evidence that is questionable and inconclusive.

In addition to the Key West–type specimen, there is one at MCZ, Harvard, collected from Sand Key by Archie Carr in 1941. What may be the oldest preserved specimen in any museum from the Keys is a fine Kemp's ridley at MCZ taken in the Tortugas, 14 April 1858, by a Captain Harrison.

References

Carr, A. 1956. *The Windward Road*. New York: Alfred Knopf (reissued by Univer- sity of Florida Presses, Gainesville).

Garman, S. 1880. On certain species of Chelonioidae. *Bulletin of the Museum of Comparative Zoology* 6(6): 123–26.

Hildebrand, H. H. 1982. A historical review of the status of sea turtle populations in the western Gulf of Mexico. In *Biology and Conservation of Sea Turtles*, K. Bjorndal, ed., pp. 447–53.

Lazell, J. 1980. New England waters: critical habitat for marine turtles. *Copeia* 1980(2): 290–95.

Lutcavage, M., and J. A. Musick. 1985. Aspects of the biology of sea turtles in Virginia. *Copeia* 1985(2): 449–56.

LEATHERBACK
(DERMOCHELYS CORIACEA)

The Greatest Reptile.

JAMES LAZELL, 1976

The leatherback, also called luth or trunk turtle, is reported to attain a weight of a ton and a total length (head, body, and tail) of over ten feet (over three meters). Only the largest crocodiles could approach a big leatherback in mass. I believe leatherbacks get heavier than any crocodile because they average much heavier than most. Accurate weights of the giant reptiles are extremely hard to find in scientific literature.

Leatherbacks are dark brown to near-black with rose, purple, or whitish splotches on chin, throat, and underside. As their name implies, they lack a bony shell fused to the ribs, present in all other turtles. The shell they have is a rind-like covering quite like the fibrous blubber of a whale except that it contains a mosaic of tiny bones called ossicles. Seven roughly parallel keels are on the back and sides.

Today leatherbacks are scarce in Keys waters, but once they nested here in good numbers. Major nesting beaches are in Cayenne and Surinam, on many Antillean islands, and scattered nesting occurs in Florida and Georgia. Females are believed to lay only one clutch, averaging less than 100 eggs, per season. These hatch in 60 to 68 days. Initial nesting may take place in spring, and it is possible that females do lay subsequent clutches at different sites as they migrate northward with the summer. Females may nest two or more years in a row, or skip a year, or nest once in a lifetime. The hatchlings vanish and must somehow, somewhere survive quite a few "lost" years. We know virtually nothing about them until they reappear as big animals.

The adults are certainly the most migratory of all reptiles. They regularly travel north—even into Arctic waters—in the summer months to get their favorite food: jellyfishes. To survive in cold water, the leatherback generates and retains heat like a mammal. Not only does the whale-like blubber provide superb insulation, but the turtle's flippers house a countercurrent heat exchanging circulatory system like that which functions in duck feet and beaver tails. The initial work on thermoregulation and warm-bloodedness in the leatherback was done right in Key West by Dr. Wayne Frair of Kings College, New York, on a turtle obtained for the purpose by Bill Ford, then of the turtle kraals. However, I have seen no museum specimens from the Keys.

References

Greer, A., J. Lazell, and R. Wright. 1973. Anatomical evidence for a countercurrent heat exchanger in the leatherback turtle (*Dermochelys coriacea*). *Nature* 244: 181.

Lazell, J. D. 1976. *This Broken Archipelago*. New York: Quadrangle, New York Times Book Co.

ALLIGATOR
(ALLIGATOR MISSISSIPPIENSIS)

Alligators will *bite the hand that feeds them.*

MICHAEL J. FOGARTY, 1978 (IN MCDIARMID)

The spectacular recovery of the alligator from the brink of extinction to genuine abundance is a monument to the success of effective wildlife management on a par with the similar recoveries of wood duck, wild turkey, and bison. Wildlife management can work; exploited species can recover and sustain a carefully measured level of renewed exploitation. Those who claim that conservationists just want to stop everyone else from harvesting economically valuable wildlife are as wrong as the bambiphiles who would have no one ever take anything. Speaking for myself, I simply want to be sure every species (or subspecies, or evolutionarily divergent population) remains healthy *as a population*. Then we can use them for whatever beneficial purpose we want.

I am not enamored of alligator leather, but I do go for the steaks. Alligators make fine zoo animals; every zoo on Earth ought to be able to display a genuine American alligator, or two.

Despite its recovery, there are still grave problems with delisting this species and opening up the leather and steak trade again. "Threatened by similarity of appearance" is the clause used to retain tight federal control (Anon., 1985). The problem is that alligator products cannot be readily separated from parts and pieces of their genuinely rare, genuinely endangered relatives. Harvest and distribution of alligator products must be closely monitored, and clever methods must be used to ensure that only legal alligator parts enter commerce. If illegally harvested parts of rare species, like some of the cascade caimans or dwarf crocodiles, could re-enter commerce disguised as American "alligator," the rare species could be exterminated easily. If one percent of leather, by weight, sold as alligator was really a rare species, that might spell its doom.

T. A. Lewis has suggested that alligators are now able to pay for their own habitat. A sustained 'gator harvest allowed by annual license fee would produce income the state of Florida could put into wetlands acquisition. As long as enough big 'gators survive the harvest to maintain 'gator holes, this system sounds quite feasible. Although Keys alligators could not support a commercial harvest, their mainland cousins might at least buy them some habitat so they may always be with us. They are a great tourist attraction.

In the Keys alligators are not really common and, in my opinion, remain endangered. A few are said to persist on North Key Largo, but I have never seen one, or direct evidence of their presence. Duellman and Schwartz marked their range as both Upper and Lower Keys, but gave no specific locations. I believe the range is widely disjunct, with the Lower Keys populations long isolated.

The easiest place to see the best-known Key alligators is the Blue Hole on Big Pine, in National Key Deer Refuge. This was originally a fair-sized natural solution hole, filled with fresh water from the large lens underlying Big Pine. The Hole was greatly enlarged some years back to fill in a parking area and provide a larger fresh water habitat. Genuinely large 'gators now live here. They are quite fearless. I suspect it will be only a matter of time before some tourist is killed by one: the tourists often seem quite fearless too, and many, I guess, are less savvy than the 'gators. It is folly to goad, or even closely approach, a big alligator.

Lewis notes that alligators in Florida kill about one person every three years, on average. We both expect this rate to rise if people do not control their numbers and keep usurping the big reptiles' rightful homes.

Alligators are widespread in the Lower Keys and often live quite unsuspected in close proximity to people. Several live on the Stock Island golf course. They can travel through solution tunnels or chambers in the oolite and pop up where least expected. I once walked up on a little sink hole where I had seen leopard frogs, hoping to see them again. The hole looked all wrong as I peered in. It cost me a long double-take to perceive the problem: the hole was entirely filled with the head of a huge 'gator. He backed and sank slowly out of sight. There was a great churning swirl in the little hole and he was back in the oolite subway system heading for his next stop. Alligators are definitely able to navigate and find their way home over considerable distances.

Alligators clear mud and muck out of holes in the process of excavating places to live. In dry weather 'gator holes can be life-savers for other denizens of the Keys, including Key deer: they provide fresh water. However, the 'gator is an important predator on the little deer. It, along with crocodiles and sharks, may have been the only significant predators on Key deer during their speciation process, prior to the arrival of humans. Today, natural predators such as alligators are of little consequence compared to automobiles, poachers, and dogs. No predator normally endangers its prey in a natural system. (See account of Key deer and bobcat.)

Alligators show rather unreptilian behavior in many respects. They make large nests at about the summer solstice in which the female lays 30 to 50 eggs. The mother attends the nest and may actually defend it, being potentially very dangerous. The eggs hatch in about 70 days. Young alligators stay together as a rule, and stay close to their mother. They communicate with a sort of high-pitched grunt, which can bring the mother on rapidly. Baby 'gators are cute little things, about nine inches long at hatching, but their mothers can be terrible and terrifying. The record size is 19 feet, two inches (584 cm—nearly six meters). Young alligators may actually play rather like mammals, but not likely momma.

I have not seen museum specimens of alligators from the Lower Keys,

Alligator, *Alligator mississippiensis*.

Joseph J. Oliver

though there probably are some at least at FSM, Gainesville. One from Big Pine Key is in the National Museum of Natural History, Ottawa, Canada. The taxonomic status of Lower Keys 'gators is a wide-open question. I am informed by crocodilian experts that this species is geographically variable from Texas to the Everglades, north to North Carolina. The Lower Keys population is the most likely to be an undescribed subspecies. The question is wholly unstudied.

Of course, alligators swim well and are not confined to fresh water. They cannot survive prolonged periods in seawater, but one cannot assume every crocodilian seen in seawater is a crocodile. Jack Watson reported to me cases of alligators taking Key deer swimming between Keys. He never mistook an alligator for a crocodile. Fogarty (in McDiarmid, 1978) provides an excellent bibliography.

References

Anon. 1985. American alligator in Florida. *Endangered Species Technical Bulletin* 10(7): 7–8.

Jacobsen, T. 1983. Crocodilians and islands: status of the American alligator and the American crocodile in the Lower Florida Keys. *Florida Field Naturalist* 11(1): 1–24.

Lazell, J., and N. Spitzer. 1977. Apparent play behavior in an American alligator. *Copeia* 1977(1): 188.

Lewis, T. A. 1987. Searching for truth in alligator country. *National Wildlife* 25(6): 12–19.

Rodda, G. H. 1985. Navigation in juvenile alligators. *Zeitschrift fur Tierpsycologie* 68: 65–77.

CROCODILE
(CROCODYLUS ACUTUS)

That curious genius, Rafinesque, one of the most surprising and
versatile of naturalists, by some hook or crook first learned of the
existence of a crocodile in Florida.

THOMAS BARBOUR, 1923

Even today many people, even those who have been here, disbelieve the existence of crocodiles in Florida. They are really here. They are potentially our second largest reptile, attaining at least 23 feet (seven meters) and weights over a thousand pounds. No Florida specimen that large has yet been recorded, but this species is widespread in the Antilles, Central and South America—and, formerly at least, in the Bahamas. It is critically endangered throughout its range. Our Florida population may be the safest.

The genus *Crocodylus* is, or was, tropicopolitan: found virtually throughout the tropics of the world. There are about a dozen species; no one knows for sure because rational taxonomic assessments are difficult to make with such large animals so sparsely represented by actual specimens. In the Americas, three besides *C. acutus* are standardly recognized. These are the Cuban *C. rhombifer*, the Central American *C. moreleti*, and the South American *C. intermedius*. The distinction of the last from *C. acutus* is rather dubious, but the other two occur in sympatry with *C. acutus* and thus appear to be valid species. Our *C. acutus* is the only one of the four that lives primarily in salt water, although it occupies fresh lakes and rivers along the Main.

At one time our Florida populations existed north along the east coast to Lake Worth: C. J. Maynard found them at the level of Lake Harney in 1872. They were well-known to the famous Barefoot Mailman of Palm Beach. Ogden (in McDiarmid, 1978) did not believe crocodiles ever bred along the west coast, but they certainly occur today as far north as Sanibel, where there is excellent nesting habitat. Most of our survivors are restricted to northeastern Florida Bay and the Lower Keys. Ogden estimated the population contained only about 25 breeding females.

Considerable maternal care involves nest-guarding, nest-opening, transport of hatchlings to water (mother carries them in her mouth), and active, aggressive protection of the young. To be successful, the female must nest where the young can live in fresh or brackish water (15–20 ppt) or where rainfall is high so that young can drink falling rain or diluted seawater. It has been believed that brackish water was absolutely necessary for survival and growth, but Paul Moler (Florida Game and Fresh Water Fish Commission, Gainesville) tells me that young seem to do well on North Key Largo in straight seawater (35 ppt). This is a rainy region.

The Everglades fringe of mainland, the bigger Upper Keys like Largo, and a few of the biggest Bay Keys like Black Betsy provide good habitats. Most of the larger Lower Keys were ideal before the coming of hordes of humanity, but only a few are today. The best is probably Little Pine, fortunately entirely within the National Wildlife Refuge. Big Torch probably supports a nest or two; this Key has provided my only first-hand observation of *C. acutus* in the Lower Keys. Adults may go most anywhere. Excellent evidence indicates that a couple live on Bahia Honda and East Bahia Honda. Individuals have been reported to me from Cudjoe, Middle Torch, Big Pine, Big Munson, and No Name. Jack Watson actually found nests on Little Pine in the 1960s. I went back to his sites, and while I failed to find nests, I ascertained that the habitat and salinities were ideal.

Females are said to lay twenty to eighty eggs in April or May. These are reported to hatch in about 100 days. The hatchlings are about 9 inches (23 cm) long; they grow rapidly and may exceed 3 feet (a meter) in two years. My remarks concerning the potential danger to humans from adult alligators apply doubly to crocodiles—especially nesting mothers. Alligators are apt to be slow and sluggish, but crocodiles are agile, amazingly fast, and active when aroused. They can leap and run—even when quite large. I began my career as an island zoologist hunting young crocodiles in Jamaica for the Philadelphia Zoo. I developed more than a healthy respect for the adults: I am very much afraid of them. I will go catch a 10-foot alligator, but you will have a hard time getting me near a 10-foot crocodile—at least in the water or in a small boat.

There are museum specimens of crocodiles from Big Pine and Key Largo, at least, but no detailed study of geographic variation in this species has yet been made.

References

Barbour, T. 1923. The crocodile in Florida. *Occasional Papers, Museum of Zoology, University of Michigan* 131: 1–6.

Dunson, W. A. 1982. Salinity relations of crocodiles in Florida Bay. *Copeia* 1982(2): 374–85.

Lutz, P. L., and A. Dunbar-Cooper. 1984. The nest environment of the American crocodile. *Copeia* 1984(1): 153–61.

Mazzoti, F. J., and W. A. Dunson. 1984. Adaptations of *Crocodylus acutus* and

Alligator for life in saline water. *Comparative Biochemistry and Physiology* 79A(4): 641–46.

Ogden, J. C. 1978. Status and nesting biology of the American crocodile (Reptilia, Crocodilidae) in Florida. *Journal of Herpetology* 12(2): 183–96.

SPADEFOOT
(SCHAPHIOPUS HOLBROOKI)

Distributed throughout Key West; commonest in vacant lots, especially on the south end of the island.

ARCHIE CARR, 1940

This appears to be the least known and most mysterious vertebrate animal in the Florida Keys. I have never encountered it. The Key West population was described by Samuel Garman in 1884 as a distinct form, *albus*. It was characterized by very light color: broad, irregular near-white areas forming a roughly lyrate dorsal pattern on a ground color of olive brown. Spadefoots from other parts of Florida may be just as pallid, and Garman's *albus* is not today recognized as valid. However, Wright and Wright (1949, and in earlier editions) maintained the subspecies and argued that it has a narrower interorbital region: the top of the skull is narrow between the eyes. They claimed that interorbital distance was contained in snout-vent length 9.3 to 10 times in *albus*, but only 6.8 to 8.5 times in nominate *holbrooki*. Dr. William Duellman (1955) did not find proportional differences, but did find a major difference in size, as Garman reported. Because Keys individuals are smaller in both length and interorbital width, the data need to be transformed. If one subtracts interorbital width from a constant, like ten, the difference as a proportion of snout-vent length does separate *albus* from mainland specimens.

Wright and Wright record this species from "Caribee Colony on Matecumbe Key." That is Upper Matecumbe. Duellman and Schwartz list no other localities, and I know of none. It is difficult for me to imagine where in today's Key West even a few might survive; possibly around the airport. It is equally difficult for me to imagine that these peculiar frogs did not also occur on others of the Lower Keys: Boca Chica, Sugarloaf, Cudjoe—all the way to Big Pine. I have never seen or heard them.

Spadefoots are fossorial creatures: burrowers. They live their lives largely out of our sight in sand, loam, and humus. They come up sporadically and sometimes at long intervals to court and breed—in a veritable frenzy of short-lived activity. Their call is a squawk. Their tadpoles are carnivorous, feeding on insects, larvae, and each other. They hatch in two days or less and metamorphose into tiny miniatures of their parents in about two weeks. Spadefoots are ideally suited to exploiting arid zones where fresh water exists for only brief periods. Their breeding activity could occur at any time of year and is triggered by heavy rains.

A big mainland adult is three inches, snout to vent (7.3 cm; Smith, 1978, gives 9 cm). Duellman did not give maxima, but found Key West specimens averaged two inches (5 cm). Females are mature and gravid at 4.3 cm. Unlike most frogs (or toads), females are no bigger than males. The tadpole is bronzy brown and a little over an inch (2.8 cm) prior to metamorphosis.

This species exemplifies the near-total neglect of the Keys amphibian fauna by the biological community. While virtually every species of snake (even if abundant and not differentiated) disjunct in the Lower Keys has made it into the *Rare and Endangered* book (McDiarmid, 1978), not one Lower Keys amphibian has. Our spadefoot may well be—or have been—a distinct form. It may now be extinct, but I doubt it.

This species was named for J. E. Holbrook, the father of American herpetology. Between 1836 and 1842 he produced two editions of *North American Herpetology,* running to nine volumes. Wayne Hanley provides an excellent account of the man, the Audubon analogue of unloved animals.

References

Duellman, W. E. 1955. Systematic status of the Key West spadefoot toad, *Scaphiopus holbrooki albus. Copeia* 1955(2): 141–43.

Garman, S. 1884. The North American reptiles and batrachians. *Bulletin of the Essex Institute* (Massachusetts) 16: 3–64.

Hanley, W. 1977. *Natural History in America.* New York: Quadrangle, New York Times Book Co.

NARROW-MOUTH FROG

(GASTROPHRYNE CAROLINENSIS)

One would never find it during the day if he did not hear it bleat during a rainstorm or cloudy weather, or if he did not unearth it from beneath its cover.

WRIGHT AND WRIGHT, 1949

This is a fossorial species. Its breeding activity is triggered by heavy rains, at which times the Wrights note, "the din is incredible." One can still hear very sizable choruses in the Keys, especially on Cudjoe Key where fresh marshes are still fair-sized. Breeding is said to occur between May and September for the species (Wright and Wright), but in the Lower Keys it might occur in any month of the year. I have heard choruses in May and June.

These are appealing little creatures. They are round, rather flat, and short-legged with small heads and pointy noses. They get to be an inch-and-a-half long (3.8 cm); females average larger than males. They are shades of brown; in the Lower Keys some may be rich reddish-brown (the "Key West phase" of field guides). They are renowned for eating termites and ants.

Females are said to lay about 850 eggs a season. These may be in a single surface film, or in "packets" of ten to 90 eggs. The floating eggs look like a mosaic because each abuts the others polygonally, as fairly straight-edged pentagons or hexagons. The tadpoles are about an inch (2.5–2.7 cm) and metamorphose in about three weeks.

I turn these little fellows up occasionally at Snake Acres on Middle Torch during my regular 100 m^2 cover search on two hectares. I have also found them on Little Torch, Summerland, Cudjoe, Sugarloaf, and as recently as 1983 around the airport on Key West. Duellman and Schwartz also record this species from Stock Island, Big Pine, Lignumvitae, at least one of the Matecumbes, and Key Largo.

The inheritance of color pattern, especially if it could be correlated to environmental or geographic variables, would make a fascinating and illuminating subject of study for a geneticist. As vertebrates go, their generation times are short and their fecundity high.

SOUTHERN TOAD
(BUFO TERRESTRIS)

This toad is truly an alert, pert animal.

WRIGHT AND WRIGHT, 1949

It is hard to mistake a toad for anything else, and this one is a classic. However, there are two other species of *Bufo*—true toads—in the Keys, and identification can be difficult. This is our middle-sized toad, attaining nearly 4½ inches (11.3 cm): females are bigger than males when adult. It is warty and patterned in brown with more or less reddish, grey, and near-black. Males have dark throats that inflate, balloon-like, when they make their trilling, droning calls.

On the top of the head are strong crests. They flank the interorbital area, medial to each eye, and turn abruptly outward, L-like, to border the rear of each eye. At the angle, behind the eyes, these crests make an upraised, knob-like prominence. At their outer ends, over the eardrums, the crests hook back to approach or touch the large, spongy-looking paratoid glands on the shoulders. The paratoids of this species are quite conspicuous but little, if any, larger than the eyeball. The other toads lack knob-like prominences at their crest angles and have either much smaller or larger paratoid glands. Baby toads are the most difficult to identify because features like crests and skin glands are weakly developed.

I have seen tiny metamorphs, but not tadpoles, in the pinelands of northern Big Pine in early June (1977). I thought a toad hopping across a street in Key West back in June 1977 was this species, but I did not collect it and now wonder if it was not just a young marine toad. Duellman and Schwartz record this species additionally from Cudjoe and Sugarloaf. Thus, the Lower Keys populations appear widely disjunct from mainland toads. Their taxonomic status is unstudied and their continued existence in the Lower Keys is extremely precarious.

Bufo seem especially susceptible to chlorinated hydrocarbon insecticides, including those like Malathion and Baygon still in use today. We heard no toads calling in May and June 1985, when we made a quick survey of amphibian breeding choruses in the Lower Keys.

These toads probably still survive on uninhabited Keys, like Little Pine. If spraying were terminated on National Wildlife Refuge lands, where it constitutes a direct abnegation of the very purpose of a wildlife refuge, stocks could either build up again naturally or—if extirpated—be reestablished.

These toads may breed in any month of the year. Breeding is triggered by heavy rains, as in most of our frog species. A female may lay a string of up to three thousand eggs. The tadpoles hatch in two to four days; they are jet-black little blimpoids with skinny tails. They take at least four weeks to metamorphose, so my Big Pine individuals probably represented a mating about the first of May.

OAK TOAD
(BUFO QUERCICUS)

Apparently it takes more rain to bring them out in full force than any other of our frogs except Scaphiopus.

ARCHIE CARR, 1940

This native species seems rare. Like the southern toad, its range appears disjunct: the Lower Keys populations separated widely from mainlanders. On the mainland oak toads are abundant. Carr wrote in 1940: ". . . I believe it would be possible to cross Central Florida by car and at no time be in a situation where the calls of *quercicus* were not audible." I have seen it twice, in 1974 and 1975, on Big Pine Key. One of the two habitats, close to Route 1, has now been destroyed by filling for a parking lot. The only other record known to me is Duellman and Schwartz's from Boca Chica.

The name of this species is a mystery to me. It probably connotes the dominant vegetation at the type-locality, far away on the mainland. This is a dwarf species, reaching an inch-and-a-quarter (3.3 cm). Females average bigger than males. Most any toad may have a light middorsal stripe, but it is especially prominent in *B. quercicus*. The appearance is velvety, patterned in shades of dark and light brown to almost black. There are tiny, red, tubercular warts. The male's throat is white, sometimes speckled with slaty-blackish. The cranial crests are weak and quite without knobs. The paratoids are not as conspicuous as in either of our other species.

The call is the best identifying character. It is a very high-pitched whistle described by the Wrights as "very birdlike" and "the most unfroglike note I ever heard" (the Wrights often write in the first person; they seem to have genuinely thought of themselves as one). It is said it can be heard over 200 meters—"a terrible din." I would love to hear them in the Keys, but I never have.

The Wrights report that local people on the mainland who were regarded as good and accurate observers of nature most of the time devoutly believed the call of this toad was made by the black snake, *Coluber constrictor.* This is uncannily like the belief one encounters virtually throughout the West Indies regarding the whistling notes of some *Eleutherodactylus* frogs (see greenhouse frog, below). I have met any number of people, some well-educated and excellent naturalists, who believe the whistling noise is made by a reptile—usually a lizard ("Z'andolie," see *Anolis,* above), but sometimes a snake. This story leads right up to the similarly widespread one about huge snakes with combs on their heads who crow like roosters so as to attract hens to their doom. (Now, I've seen a lot of roosters chasing hens but I regard a hen homing in on a crowing rooster the same way I regard snakes that chase me: no evidence in my lifetime.)

Of course it is incredibly easy to disprove the whistling snake (or lizard) belief. When I was a boy and first heard this tale (in Dominica, the Windward Islands) I led my good friend—on whose every word concerning animals and nature I hung—out and showed him the frog, in the flashlight beam, obviously making the noise in question. Well, he said, maybe that *one* was indeed a frog, but not all the others. That night I learned a lesson about the nature of evidence and human belief. It was a hard lesson with repercussions so profoundly sad, for our own species and every other we contact, that I have never yet readjusted to it.

Oak toads are said to breed from April to September; I doubt they are so restricted in the Lower Keys. The eggs are laid in short strings, or bars, of up to half-a-dozen. I have no total count per female, but it must be hundreds. I do not know how rapidly they hatch or how long the tadpoles take to reach metamorphosis, but it must be a few days and a few weeks, respectively. At metamorphosis the tadpoles are small—about a quarter inch (7.8 mm). I have never seen the tadpoles, but the Wrights' description indicates they are quite distinctive: "The small tadpole is grayish with six or seven black saddles on the musculature, and with heavily marked upper tail crest, and the venter is one mass of pale purplish vinaceous." Vinaceous means wine-colored; it tormented me for years so I'll save you the trouble of looking it up. Birders use it all the time.

I can scarcely imagine that one or another of the disjunct Lower Keys amphibians has not evolved into a novel form. It seems that, apart from Samuel Garman's spadefoot, no one has really considered the question, although the Wrights did postulate it. The little oak toad might be the best prospect for marked speciation, if it survives here at all.

MARINE TOAD
(Bufo marinus)

...the paratoid glands and certain other modified warty glands on the backs of these toads secrete a milky fluid that is highly toxic if ingested, rubbed into the eyes, or brought in contact with mucous membranes.

Sean McKeown, 1978

This is our giant toad. Big females from the Key West–Stock Island population attain about seven inches snout-to-vent (18 cm), but in their native haunts in South America they are said to attain 23.8 cm. Males average smaller. These toads have been widely introduced throughout the tropical world, ostensibly to control large insects such as beetles and roaches. They will eat most anything, from dog food to fruit.

Carr did not know them from Florida before 1940, but Duellman and Schwartz had them on the mainland, where they are now abundant in Dade County, by 1958. Krakauer stated they reached Miami prior to May 1955, but they were not known from the Keys in 1968. So far, they seem restricted to Key West and Stock Island in the Lower Keys, but they will surely spread. They are native in the United States only in extreme South Texas.

As with other *Bufo*, the coloration is basically shades of brown. The cranial crests are prominent, but form the border of a broad, concave, dishlike top of the head. There are no knobs. The paratoids are huge, at least three times the size of the eyeball, and extend well back along the sides of the body. Krakauer compared the diet of this species to that of the native *Bufo terrestris* (above). Marine toads in Dade County ate large numbers of beetles and mollusks (slugs and land snails), and also considerable vegetable matter such as fruit. Southern toads ate fewer and smaller beetles, and virtually no mollusks or fruit.

Marine toads may breed in any month of the year. The males emit a resonant "deep booming trill" (the Wrights). The females lay strings of hundreds of eggs. The tadpoles may hatch in as few as 68 hours and metamorphose in 45 days or less (the Wrights). Tadpoles did much better in 15 percent seawater (*ca* 5.4 ppt salt) than in tap water in experiments conducted by C. A. Ely in 1945 (quoted by the Wrights). During the rainy autumn of 1985 marine toads bred successfully in little more than puddles in Key West. From one quite brackish site, Whistling Duck Pond near the southern end of White Street, swarming black tadpoles and dozens of tiny metamorphs were observed by A. B. Ford on 19 October 1985; she did collect voucher specimens, now in MCZ, Harvard.

Carpenter and Gillingham studied marine toads where they have been "naturalized" in Puerto Rico. They found adults had "activity centers" (they eschew the term home range) of about 390 square meters (varying from a female with 167 m^2 to a male with 862 m^2). These centers were often at some remove from permanent fresh water and the toads had to travel from 15 to 165 meters (average 76 m) to get to water. The toads seemed to really know their surrounds, because they would shift to another water hole if the one they normally used dried up. In dry weather, a fresh water source was critical to their survival.

Because it is an introduced exotic, I should deplore this monster in our midst. I have yet to hear compelling evidence that it is bad for our native species; most of what it eats are other introduced exotics. It is surely a nuisance when abundant, and genuinely toxic if not handled with care (why handle one at all?). As a beneficial or detrimental member of our fauna, the jury is still out on this fellow.

Marine toad, *Bufo marinus*. Key West.

Joseph J. Oliver

References

Carpenter, C. C., and J. C. Gillingham. 1987. Water hole fidelity in the marine toad, *Bufo marinus. Journal of Herpetology* 21(2): 158–61.

Krakauer, T. 1968. The ecology of a neotropical toad, *Bufo marinus*, in South Florida. *Herpetologica* 24(3): 214–21.

McKeown, S. 1978. *Hawaiian Reptiles and Amphibians*. Honolulu: Oriental Publishing.

Wilson, L. D., and L. Porras. 1983. The ecological impact of man on the south Florida herpetofauna. *University of Kansas Special Publication* 9: 1–89.

GREENHOUSE FROG
(ELEUTHERODACTYLUS PLANIROSTRIS)

The genus Eleutherodactylus *is noteworthy for the richness and diversity of its species and for the problems of distribution and speciation which it poses.*

COLEMAN J. GOIN, 1947

I first met Dr. Coleman Goin and his wife, Olive, in the Blue Mountains of Jamaica in 1957. I went there especially to meet him and learn about his work on frog genetics first hand. Goin pioneered efforts to understand genetics and patterns of heredity in vertebrate animals. All too few other researchers have taken up this sort of slow and tedious work. Instead, fast-breeding, incredibly prolific organisms like fruit flies and colon bacteria have dominated the laboratories and investigations of geneticists. For better or worse, most genetic theory and practice today are based on these organisms, rather than on vertebrates far more closely related to us.

Our *Eleutherodactylus* come in at least three color patterns: striped, mottled, and a nearly unicolor mutant. Goin determined that a classic Mendelian inheritance pattern controlled striped and mottled, with striped as dominant and mottled as recessive. The unicolor mutant (and some odd speckled Bahama specimens) were so rare Goin could not determine their heredity. Simply put, a mottled frog must inherit the recessive gene for mottled color from *both* parents. A frog that inherits a dominant gene for striped pattern from *either* parent will appear striped even if it carries the gene for mottled inherited from the other parent. A frog with one gene for striped and the other gene for mottled is called a *heterozygote*. Since it is striped, we cannot visibly tell it from a frog that has two genes for striped—one from each parent, called a *homozygote*. We can tell them apart by breeding them and seeing what ratios of offspring they produce.

Breeding two striped frogs has two possible outcomes. We may get a few mottled frogs and a lot of striped ones, in approximately a three to one ratio, which indicates *both* striped parents were heterozygotes. Each had a fifty-fifty chance of giving a gene for mottled to each offspring. Those offspring that got a gene from each parent for mottled come out mottled, but those that inherited a gene for striped from *either* parent are just as striped as the ones that got genes for striped from *both* parents. Thus, about two-thirds of the striped frogs—about one-half of the total production of offspring—will be heterozygotes. The ratio of offspring by genetic constitution will approximate 1:2:1, as homozygous striped; heterozygous striped; homozygous mottled, respectively. That is the genotypic ratio: what they really are, not how they look. The way they look approximates a 3:1 ratio, striped to mottled. If *either* striped parent is homozygous, *all* the offspring will be striped.

The only way to be certain of the genetic constitution—genotype—of a striped frog is to cross it with the *double recessive,* a mottled frog. If offspring are 100 percent striped the striped frog was homozygous. If offspring are about half-and-half, striped and mottled, the striped frog was heterozygous. You can figure out the genotypes of the offspring in each case.

Things are never so simple in nature as I have described them above. There are modifier genes, blending effects, and those odd, rare phenotypes that just do not quite fit. When I met the Goins thirty years ago they were studying *E. nubicola,* which has at least *four* distinct phenotypes that are genotypically controlled. Even if you are not at all interested in the color patterns of little frogs, you can perceive the significance of such studies. Increasingly we need to know patterns of heredity in our own species. Medical art makes it possible for many people carrying genes once quite lethal—like the one for childhood diabetes—to go right on living and reproducing. We cannot perform simple breeding experiments with humans, even if we suspect from eyeballing phenotypic ratios that we know the genotypic pattern of a disease. We are every day more frequently faced by the fearful specter of genetic disease affecting more and more of us. We suspect, for example, that one form of schizophrenia results from a single dominant gene that causes a chemical dysfunction in the brain. If this is true, the children of such schizophrenics have a fifty-fifty chance of being schizophrenics even if their other parent is quite normal. The children of two heterozygotes for such a condition have only one chance in four of being normal.

While the prospects for genetic engineering make it likely that we can correct phenotypic disorders in individuals, there is as yet no real method for changing the genotype of any human. I, for one, can readily foresee the day when every human on Earth will be an artificial product of genetic engineering and biochemical manipulation: doctored phenotypes based on hopelessly defective genotypes. If you think the cost of medicine is high now, wait a few generations.

The alternative, discussed by Kelves in his book cited below, is a knowledgeable and rational approach to genetics, based on a vast knowledge of patterns of heredity. One day we may have more to thank workers like Coley Goin for than just elucidating the natural history of pretty little frogs.

Our *Eleutherodactylus* has been presumed to have been introduced, but like the Cuban treefrog and some *Sphaerodactylus* geckos it may as well have reached the Keys quite naturally by over-water waif dispersal from Cuba long before humans came that way themselves. First scientific knowledge of our populations came in 1863, when Edward Drinker Cope of Philadelphia's Academy of Natural Sciences noted them at Key West. As Carr (1940) points out, they had been there for a long, long time before that.

They have certainly spread rapidly since then. They were known from Miami by 1899, Brevard County by 1910, Gainesville by 1933, and Jacksonville by 1943. They seem not to have moved farther north, but occur as far west as Tallahassee today. Excellent evidence indicates that their spread is augmented by humans. Apparently their eggs are transported in soil, humus, and potted plants. Unlike most of our other amphibians, these *Eleutherodactylus* lay their eggs in moist soil, under cover. The eggs undergo direct development inside a fluid envelope contained within tough jelly layers surrounding each egg.

Females lay clutches of three to 26 eggs (average 16) during the rainy months of the year—at least April through September in the Keys. Development is rapid and seems to be speeded by warm weather. It takes up to 20 days at Gainesville, but probably fewer in the Keys. Warm clutches can hatch in 13 days. The color pattern can be seen through the egg while the young still bear tails. The tiny neonates, about four millimeters at hatching, have a horny egg tooth for cutting out of their jelly egg shells. It seems that 100 percent humidity is necessary for successful hatching. Although I have many data on these most common of all Keys frogs from Snake Acres on Middle Torch Key, I have never yet found a nest. William Ford of Key West found neonates extremely abundant there from 27 May to 2 June 1985.

I have never made an effort to determine pattern frequencies (phenotypic ratios) in the Keys, but it seems to me that some pattern types besides simple striped and mottled occur. Many individuals are rich reddish, while most are shades of brown. A big adult is an inch-and-a-quarter (3.2 cm) snout to vent. The male's voice is a high, short whistle. I would welcome a serious study of these frogs in the Keys and will provide the theater of operations at Snake Acres.

This is one of our few species for which we have detailed information on the diet. Goin found 96 kinds of insects and other invertebrate prey in the stomachs of 23 individuals. Broadly grouped, about 70 percent of the diet was ants, small beetles, and roaches. These little frogs are highly beneficial indeed.

I have examined specimens from Key West, Stock Island, Cudjoe, Summerland, Middle Torch, Little Torch, Big Pine, No Name, Upper Matecumbe, and Key Largo. Duellman and Schwartz add Sugarloaf.

References

Goin, C. J. 1947. Studies on the life history of *Eleutherodactylus ricordii planirostris* (Cope) in Florida. *University of Florida Studies, Biological Science Series* 4(2): 1–66.

Kelves, D. J. 1985. *Genetics and the Uses of Human Heredity.* New York: Alfred A. Knopf.

GREEN TREEFROG
(HYLA CINEREA)

. . . an aristocratic looking tree toad, with its long, slender figure of the brightest green, edged on each side with a band of pale gold or silvery white.

R. F. DECKERT, 1915 (IN WRIGHT AND WRIGHT)

I have watched them stalk insects . . . ; the frog's extremely long hind legs render the stealthy approach to the prey a somewhat ridiculous spectacle.

ARCHIE CARR, 1940

Elegant is my word for these little creatures. They attain two-and-a-half inches (6.4 cm) snout-vent length, but with their legs extended they are nearly three times that long. Females get bigger than males. In addition to their bright green with pale, bold trim, they usually have bright yellow spots, especially in the shoulder region. They are superb climbers, with the great, enlarged toe-tip disks that provide adhesion surfaces even for walking up glass.

The call of the male is a monosyllabic, metallic note often likened to a bell. To me, it sounds like "quint." It is said (Wright and Wright) their breeding season is from April to August. I have heard choruses in May and June on Big Pine Key and Cudjoe. One chorus began with a few calling at 9:30 p.m. By 10:30 the numbers had gotten to their maximum; by 11:30 they had all gone silent. At 12:10 a.m. they began again and kept going till 12:50. After that I fell asleep, but a day later they were going full blast at 6:00 a.m., 40 minutes before sunrise. While the weather throughout this period was rainy, the choruses did not seem to be triggered by individual downpours.

Calling may begin well away from water; many authorities, in fact, recognize a distinct "rain call," as opposed to the mating call. Calling males tend to move closer to the potential breeding ponds and marshes; I cannot separate two sorts of noises. Females lay about a thousand eggs at a time, attaching them to submerged or floating vegetation in small packets. I do not know how rapidly they hatch.

Tadpoles are described as green with the light side stripe well indicated from snout to eye; they exceed an inch-and-a-half (4 cm) prior to metamorphosis. They metamorphose in about 60 days (Wright and Wright). I have found them breeding in slightly brackish water—up to 5 ppt—both in the Keys and on the Outer Banks of North Carolina.

J. D. Kilby analyzed the stomach contents of 497 specimens wild-caught at Gainesville, Florida. He found them to be virtually entirely insectivorous. Worms and crustaceans were absent from their diets, and they ate only a few small snails or slugs. Arthur Freed, also working at Gainesville, found wild individuals actively sought worm-like moving objects. Thus their diet was heavy on moth caterpillars and beetle grubs (not genuine annelid worms, however). In captivity Freed got different results: they preferred house flies to mosquitoes.

Freed found that some individual green treefrogs basked in the sun. This elevated body temperature, of course, and increased the rate of food digestion, thus increasing growth rate. Do our green treefrogs in the Keys bask? Does a generally higher ambient temperature increase their growth rate over mainlanders?

Preserved specimens are at MCZ, Harvard, from Cudjoe, Big Pine, and Upper Matecumbe. I have never heard this species on Middle Torch, for example, even on the same nights as big choruses on Cudjoe. While I feel sure green treefrogs are more widespread than my data indicate, I have no evidence.

The mosquito ditching and filling that have destroyed so much of our fresh water marshland in the Keys have undoubtedly greatly restricted this highly beneficial insectivore. It is one more in the constellation of species from Key deer to ringless ringneck snake that need the intact fresh water lens. It is unstudied in terms of geographic variation and possible endemicity.

References

Freed, A. N. 1980. Prey selection and feeding behavior of the green treefrog (*Hyla cinerea*). *Ecology* 61(3): 461–65.

————. 1980. An adaptive advantage to basking behavior in an anuran amphibian. *Physiological Zoology* 53(4): 433–44.

————. 1982. A treefrog's menu: selection for an evening's meal. *Oecologia*, Berlin, 53: 20–26.

Gerhardt, H. C., and G. M. Klump. 1988. Masking of acoustic signals by the chorus background noise in the green treefrog: a limitation on mate choice. *Animal Behaviour* 36(4): 1247–49. Describes and cites studies of vocalization and mating strategies.

Kilby, J. D. 1945. A biological analysis of the food and feeding habits of two frogs, *Hyla cinerea cinerea* and *Rana pipiens spenocephala*. *Quarterly Journal of the Florida Academy of Sciences* 8(1): 71–104.

SQUIRREL TREEFROG
(HYLA SQUIRELLA)

The colors of this animal are even more changeable than in any species with which I am acquainted.

J. E. HOLBROOK, 1842

Our smallest treefrog, this species attains little more than an inch-and-a-half (4.1 cm): the sexes are nearly equal in size, with the record to a female. The legs are proportionately shorter than in the green treefrog; the appearance is much plumper.

These frogs may be bright green, yellow-green, brown, or mottled and spotted. One individual may show the entire repertoire at different times, and the presence, absence, and arrangement of markings varies from one individual to another, making this a hard frog to identify. I cannot distinguish these from juvenile Cuban treefrogs (below), but the proportions and lack of the sharply defined side stripe separate them from green treefrogs.

The voice is the best recognition character. It sounds like the scolding call of a grey squirrel and is frequently given high up and far from water— a classic rain song. It is variously described as a harsh or rasping trill and may be quite ventriloquistic. I have heard it only on Summerland Key, in dense, fairly high hammock near the fresh water pond near the south end. Not a single frog of any sort could be heard at that pond: it was silent as a grave.

Next to the spadefoot, this is for me our rarest and least-known frog. Two specimens at MCZ, Harvard, are labeled simply "Florida Keys." Dr. Henry Fowler of terrapin renown recorded these on Key Vaca and Boca Chica, back in 1906. His specimens should be in the Academy of Natural Sciences in Philadelphia. Carr (1940) regards this as the commonest Florida treefrog, and Duellman andSchwartz provide a fine list of Keys: not only Boca Chica and Key Vaca, but Key West, Cudjoe, Little Torch, Big Pine, Long Key, Upper Matecumbe, and Lignumvitae. Had we been able to catch one, our Summerland callers of 26 May, 1985 would have been a new island record.

Squirrel treefrogs are said to breed from April to August. Their eggs are laid singly on the bottom of shallow fresh to slightly brackish ponds and marshes; the total complement per female is given by the Wrights as 950. I do not know how long they take to hatch, but they are described as "citrine drab" with a whip-like tail tip. They reach an inch-and-a-quarter (3.2 cm) prior to metamorphosis in 40 to 50 days (Wright and Wright).

Throughout most of its range this is the species most often associated with humans. It is truly edificarian, climbing on glass and screens at night and seeking shelter behind shutters and under roofing by day. In the Keys this niche has been usurped by the huge Cuban treefrog (below) and the squirrel treefrog may have been reduced to rarity through competition or outright predation. Cuban treefrogs love to eat other frogs. Like our other natives, this seems to be a wholly neglected member of our fauna.

Wilfred Neill, on the mainland, located two daytime retreats or homes of an adult squirrel treefrog. These were 43 feet apart; to get from one to the other the frog had to cross over ground. The frog often foraged on the buildings where these homes were located, and always stayed out all night long: ten to eleven hours. This fact suggested to Neill that the frog depended on sight to locate landmarks and literally could not find its way home in the dark. It all wants a great deal more study and experimentation to prove such a point.

Reference

Neill, W. T. 1957. Homing by a squirrel treefrog, *Hyla squirella* Latreille. *Herpetologica* 13(3): 217–18.

CUBAN TREEFROG
(HYLA SEPTENTRIONALIS)

The species may have rafted to Florida from nearby Cuba and/or the Bahamas, or it may have been introduced by man; both alternatives are feasible, and perhaps both occurred.

L. TRUEB AND M. J. TYLER, 1974

I have seen twenty-five or thirty clinging to the walls of a single cistern in Key West.

ARCHIE CARR, 1940

This species is also called the giant treefrog, although other Antillean species grow much larger. It is popularly put in a genus "*Osteopilus*" (see Trueb and Tyler), said to differ from *Hyla* in minor aspects of musculature and (especially) in having the skin fused to the top of the skull. This latter character sounds impressive until you examine specimens. It is undetectable in young *septentrionalis,* rendering the species hard to identify (let alone the "genus"); it is widespread to one degree or another in many other species of *Hyla* from other parts of the world. The hylid frogs have been divided into a plethora of morphologically indistinguishable and often interbreeding "genera" on the basis of inconsistent characters and far-out theoretical concepts ("molecular clocks," something like the Holy Grail). We need fewer, better genera; not more.

These tree frogs attain the size of bullfrogs (*Rana catesbeiana;* not resident in the Keys), and are even called that by some people. They can reach five-and-a-half inches (14 cm) snout to vent. Their legs are long and their toe disks huge. They climb all over everything.

Coloration is extremely variable, involving shades of green, yellow, and brown—reaching lustrously bronzed platinum blonde, silver, or ash-

Joseph J. Oliver

Cuban treefrog, *Hyla septentrionalis*.

white in some big adults. They begin life smooth of skin, like our other treefrogs, but big, old fellows are so warty one can hardly imagine them the same species. I think the Cuban treefrog is one of the most delightful and wonderful animals alive.

It certainly is successful. Before World War II, Carr knew it only from Key West, where as he pointed out the oldest residents had always known it. The species was actually far more widespread, however, even then, for the Wrights recorded it at Upper Matecumbe in 1934. Duellman and Schwartz by 1958 had it from Stock Island, Big Pine, Key Vaca, both Upper and Lower Matecumbe, and Key Largo. Today it is expanding its range on the mainland. It is abundant at Snake Acres on Middle Torch, where I have copious activity notes on individuals from January to March. I have collected series of tadpoles, metamorphs, juveniles, and adults on Little Torch and Cudjoe and sent them to the Chengdu Institute of Zoology in Sichuan, China.

I have found them most often breeding in January and February, but believe they must breed year 'round. However, we recorded none calling in our quick chorus survey in May 1985. (Juveniles and adults were everywhere at that time, however silent.) The Wrights give June to September.

The calls are highly variable in pitch, probably because males span such a great range of sizes at sexual maturity (from at least 4.6 to 7.5 cm). The voice is likened to a rasping snarl or a snore, and sounds rather like the voice of the spadefoot—which could make real problems in the field. Breeding takes place in any available vessel of fresh water. Cisterns and ponds are favored, but buckets, outdoor aquaria, bird baths, and old tires will do for a good try. I do not know how many eggs each female lays, but likely a thousand. My own field experience, in the Antilles, is that the eggs hatch in less than a week; the tadpoles are brown and attain about an inch-and-a-half (4 cm) prior to metamorphosis; they metamorphose in about four weeks. The pretty, blue-green metamorphs I know are smooth and not at all warty as the Wrights describe. I expect there is considerable variation.

Dr. William Robertson reports seeing and hearing this species at Garden Key in the Tortugas. It is the only amphibian known from these islands and we need voucher specimens. Certainly no frog was ever mentioned or collected by the Holders, Mills, or Harrison who sampled the Tortugas fauna in the nineteenth century. I feel sure its presence at Fort Jefferson today is the result of recent human introduction, but I have no proof of that.

I am less convinced that *Hyla septentrionalis* was introduced to the Keys by humans in post-Columbian times. It may well be native in at least the Lower Keys, and have undergone population explosion and range expansion concomitant with that of humans. It is certainly an edificarian form, like the opossum and the green anole lizard. No one doubts that those species are old natives, albeit more abundant and widespread today than 500 years ago.

Reference

Trueb, L., and M. J. Tyler. 1974. Systematics and evolution of the Greater Antillean hylid frogs. *Occasional Papers of the Museum of Natural History,* University of Kansas, 24: 1–60.

LEOPARD FROG
(RANA UTRICULARIA)

*We present this rambling treatment to provoke further study—
study of all the collections plus travel experience with these very
forms. . . . We are past appraisals of alcoholic specimens alone.
Those began 75 years ago.*

ALBERT HAZEN WRIGHT AND ANNA ALLEN WRIGHT, 1957

The great leopard frog complex covers most of temperate North America.
There are many forms, morphs, subspecies, and full species. Bales and
reams have been written on the systematics of these frogs; they are among
the most ubiquitous and important—in dollars—of all wild or laboratory
animals. The Wrights alone donated more than 46 pages, about eight
percent of their entire text, to these frogs. We have now been at it more
than 110 years and we still are not through.

The disjunct populations of the Lower Keys have never been studied,
described, or clearly identified. The name I have given them, *Rana utricularia,* is at least the most currently popular name for what is at least the
most proximate mainland relative. For many years this frog was known as
"*Rana pipiens sphenocephala.*" It is locally abundant on Big Pine, all three
Torch Keys, Ramrod, Summerland, and Cudjoe. Duellman and Schwartz
record it from Key West, but at least some leopard frogs currently in
backyard ponds in Key West are said to have been bought from mainland
supply houses. I am sure this species was originally native on Key West. It
probably persists on Stock Island and likely occurs on Boca Chica, Sugarloaf, and Little Pine as well.

Our Lower Keys individuals attain enormous size for their species—to
nearly five inches (more than 12 cm) snout to vent. They are dark and
somber, often with dark bellies, as noted by the Wrights. They are tolerant
of very brackish waters, up to 10 ppt—approaching one-third the salinity
of seawater. They breed in every month of the year, especially in rainy
weather.

The males produce a true croak, quite as described by Aristophanes
some 24 centuries ago. I have recorded eight separate choruses in an
evening, beginning just before the rain at 7:00 p.m. and continuing till
after midnight, in February at Snake Acres on Middle Torch Key. I have no
specific data for the Keys, but mainland females lay their eggs in big
bunches—the total apparently uncounted. The egg masses are attached

to submerged objects like vegetation. The tadpoles grow to be three inches (7.4 cm) and develop black blotches on their tails.

Leopard frog tadpoles metamorphose in more than two but less than three months. Young individuals may be quite bright green (blue is common, too, at least on the mainland; I have not seen blue individuals in the Keys). The common name derives from the bold, blackish, oval spots. These may become obscure with age, as the animal darkens all over. Our leopard frogs seem not to stray far from marshes and ditches with fresh or brackish water, but they are rather terrestrial in habits. They are ravenous insect eaters, like most frogs, and highly beneficial.

J. D. Kilby studied the stomach contents of 443 wild-caught at Gainesville. In addition to insects, they ate lots of annelid worms, crustaceans, and scorpions. More than seven percent had eaten mollusks: slugs and snails. Nearly nine percent had eaten vertebrate animals, especially other frogs.

Dr. William Dunson, of Pennsylvania State University, has begun a study of these salt-tolerant amphibians. We can predict a great improvement in our knowledge of them soon.

Reference

Kilby, J. D. 1945. A biological analysis of the food and feeding habits of two frogs, *Hyla cinerea cinerea* and *Rana pipiens spenocephala*. *Quarterly Journal of the Florida Academy of Sciences* 8(1): 71–104.

INSECTS
AND OTHER
INVERTEBRATES

Insects are the dominant group of animals on the earth today. They far surpass all other terrestrial animals in numbers, and they occur practically everywhere.

BORROR, DELONG, AND TRIPPLEHORN, 1976

My knowledge of invertebrate life—despite its ubiquitous abundance—is paltry to the point of pathetic. This section of my work is eclectic and fragmentary. I can provide only brief glimpses into the insect and other invertebrate life of the Keys and suggest that any line of serious investigation one took up would surely produce wonderful results.

For many people the only insects of interest are the obnoxious ones that bite or sting, and interest is only in avoiding them. Those people are hardly likely to read this book, but I admit to thoroughly disliking mosquitoes, no-see-ums, and chiggers (the last are not actually insects) too. There are plenty of these in the Keys. The Monroe County Extension Service office of the University of Florida, located on Stock Island, can supply information on pest insects including those mentioned above and horticultural and agricultural pests too.

We have two common genera of mosquitoes: *Aedes,* which includes salt marsh and woodland mosquitoes, and *Culex,* the common house and garden mosquitoes that breed in any body of fresh water down to a tin can. The scarcer genus *Anopheles* is widespread in the Southeast, Antilles, and most of the tropical world. *Anopheles* are easily recognized by their upright posture: when they alight they rest at an acute angle; when they bite they elevate posteriorly to near-vertical. *Anopheles* can carry malaria; *Aedes* (at least species *aegypti*—rare or absent here) can carry yellow fever; and

217

Culex can carry encephalitis and dengue fever. Fortunately for us these diseases are not currently present in the Keys.

The no-see-ums or punkies are small flies closely related to mosquitoes. Mosquitoes are dipterans (like house flies) of the family Culicidae; no-see-ums are dipterans of the family Ceratopogonidae. No-see-um larvae live in mud or sand or shallow water and many are marine. I am not aware of any disease vectored by no-see-ums. A favorite Key repellent for no-see-ums is Avon *Skin-so-Soft* diluted in tap water.

Chiggers are arachnoids distantly related to spiders. They belong to the family Trombiculidae, and ours are probably genus *Trombicula*. It is the larvae, which are bright red (and cause red welts), that feed on us. The adults are predators on insects and insect eggs, so chiggers are not all evil. In the Orient chiggers carry scrub typhus, a potentially lethal disease. That has not been reported here. Chiggers usually climb up us from the ground, so a good soaking of insect repellent around the ankles will tend to keep them off.

There are two families of ticks, which are also arachnoids, not insects. The hard ticks, family Ixodidae, include dog ticks and deer ticks. These attach to their host and inflate with blood before dropping off. The soft ticks, family Argasidae, do not attach to their prey, but take blood meals whenever they can. They are especially fond of nesting birds and are abundant at terneries and other rookeries. I have not found ticks to be a problem for humans in the Keys.

Cockroaches, also called water bugs and palmetto bugs, cause no end of consternation among humans, but so far as I know are harmless. Roaches are classified in the order Orthoptera by me because that is how I learned the system. At least up to edition four, Borror et al. agreed. Many entomologists now put them in a separate order, Dictyoptera, but I will need some convincing. Not only do many roaches look much like crickets, some actually stridulate—"sing" by rubbing two surfaces (leg/wing; wing/wing) together. We have two families, at least, represented by big, obvious forms in the Keys.

The roach family Blattidae includes the native American genus *Periplaneta*, which has wings and flies well as an adult, and the stinking roach, *Eurycotis floridana*, which cannot fly with its very short wings. The family Blaberidae includes the Cuban roach, *Blaberus crannifer*, which is about our largest species, reaching more than two inches (50 mm). These usually live outdoors, under cover in the woods. The insecticides people use to wage war on roaches are genuinely dangerous and may cause cancer. The roaches do not vector diseases, or bite, or sting. I vastly prefer roaches to cancer, but some may disagree. Unfortunately for me, my neighbors' disagreement may cause my demise, which hardly seems fair.

Another group of orthopterans (in the broad sense) that are especially notable in the Keys are called walking sticks. These are relatives of the praying mantis, family Mantidae (which also occurs here), but lack the formidable forelegs, bulging eyes, and ravenous expressions of those carnivores. Walking sticks belong to the family Phasmatidae. Unlike mantids, which are good flyers, phasmatids are wingless, or have very short wings, and cannot fly. Our most common species is *Anisomorpha buprestoides*. It is brown and longitudinally striped. Females attain at least four inches (10 cm) head-body length, but males are much smaller. These walking sticks are often found under cover, beneath slabs of oolite, in stumps or logs, or in junk piles. Sometimes dozens or hundreds congregate together. They are not dangerous, but they do secrete a foul-smelling noxious fluid if handled. This fluid can be very painful if one gets it in the eyes. Another species has been reported to me and is very slender and bright green, like pine needles. I have yet to see it.

Most people believe coneheads are comic creatures of science fiction, but two real species live in the Keys. They are insects of order Orthoptera, as is the threatened Keys scaly cricket, *Cycloptilus irregularis*. All three are reported on by Walker (1982). The Big Pine conehead, *Belocephalus micanopy,* is known only from Big Pine, but must surely occur on No Name, and possibly other pine habitats such as occur on Upper Sugarloaf and Little Pine. Over one inch, head-body length, this species varies from green to brown, has short wings, and is flightless. The head is conical indeed and the eyes are located near the pointed top.

A second, larger species, *Belocephalus sleighti,* approaches two inches, head-body length (four centimeters). It has a large (four millimeter)

Walking stick, *Anisomorpha buprestoides*.

Joseph J. Oliver

thornlike spike projecting from its conical head, entirely extending beyond the eyes. It too is green or brown, but not confined to pinelands. It is known only from the Keys, from Sugarloaf to Plantation Keys.

Like katydids (to whose family they belong), coneheads stridulate to make songs in the evening. Their calls are described as "regular sequences of short rattly bursts" by Walker (1982). The genus *Belocephalus* is endemic to Florida; three other species occur on the mainland.

There are lots of ants in the Keys of course (Wilson, 1964), but to my knowledge no fire ants. The fire ant, *Solenopsis saevissima richteri*, was introduced along the Gulf Coast from South America several decades ago. It has spread rapidly and is a major nuisance. Its bite causes a nasty blister and real pain. We have some nippy ants in the Keys, but nothing to compare to the real thing. I do not know how we have managed to be so lucky—so far.

At least four other members of the genus *Solenopsis* occur in the Keys. One, *S. geminata*, is called the "native fire ant," but its bite is relatively mild. Wilson recorded a total of 30 ant species, of which eight (*ca* 27 percent) are "tramp" species that man has spread around the world inadvertently. Small islands seem especially prone to colonization by tramp species. Of the remaining 22 species, ten—nearly half—are of Antillean origin. This is a much higher percentage than in any vertebrate animal group. Wilson points out that most (six) of the Antillean species are tree-dwellers and that their dispersal over water is greatly facilitated by hurricanes. Trunks and branches containing living ants may carry potential colonies great distances. None of the ant species known is endemic to the Keys, but one, *Camponotus tortuganus*, was originally described from the Dry Tortugas. It occurs as far north as Lake Worth on the mainland.

Giant centipedes live in the Keys. They are rich orange, yellow, and

Giant centipede, *Scolopendra alternans*.

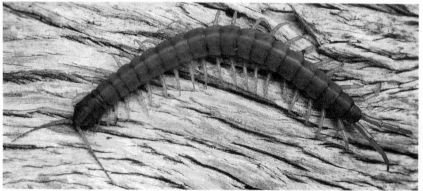

Stewart B. Peck

brown in colors, and may attain eight inches (20 cm) in length. They have paired fangs under their heads and can inflict painful bites. The caudal appendages—at the opposite end—scare people but are innocuous. The giant centipedes I have collected seem to be *Scolopendra alternans,* but *S. morsitans* and *S. subspinipes* may also occur here. The taxonomy of centipedes is scarcely studied, as is perforce their biogeography. Are these native? Are they Antillean invaders? Did they get introduced by us humans from far away places with strange-sounding names—like the house geckos of the genus *Hemidactylus?*

Centipedes are not insects. They belong to the phylum Arthropoda, along with insects, spiders and their kin, crabs, shrimp, lobsters, and barnacles (for example), but are put in the class Chilopoda. These are segmented, multilegged arthropods with one pair of legs per segment. The next class over, Diplopoda, contains the millipedes, which have two pairs of legs per segment. The word centipede means a hundred legs, but none has that many. Ours have 21 to 23 pairs. Some millipedes have a hundred legs, but their name, hyperbolically, means a thousand legs. None approach that, of course.

Another remarkable arthropod native to the Keys has made itself known to me. This is the fresh water crawfish (or crayfish, if you like), *Procambarus alleni.* It occurs in the fresh water lens of at least Big Pine Key and grows to about five inches (13 mm). I encounter it when hunting frogs at night. It is a North American species stranded in the Lower Keys by sea level rise since the Wurm glacial maximum, like leopard frog and alligator.

Recent estimates put the probable number of arthropod species on Earth at something like 30 million. Only a million or so have even been named or described, so you can understand why many scientists regard this planet as extremely little-known. The problem is to find out about the great diversity of life on Earth before human overpopulation and destructive overdevelopment wipes it out. If you decide to become an entomologist you will have plenty to do.

References

Borror, D., D. DeLong, and C. Tripplehorn. 1976. *An Introduction to the Study of Insects, Fourth Edition.* New York: Holt, Rinehart and Winston.

――――, and R. E. White. 1970. *A Field Guide to Insects.* Boston: Houghton Mifflin Co.

Evans, H. E. 1984. *Life on a Little-Known Planet.* Chicago: University of Chicago Press.

Hobbs, H. H. 1942. *The Crayfishes of Florida.* Gainesville: University of Florida.

Maxwell, L. S. 1959. *Florida Insects.* Published by the author, Tampa; sold by Florida Audubon Society, Maitland.

Walker, T. J. 1982. Order Orthoptera. In *Rare and Endangered Biota of Florida, Volume Six, Invertebrates*, R. Fanz, ed., pp. 46–48.

Wilson, E. O. 1964. The ants of the Florida Keys. *Breviora,* Museum of Comparative Zoology, Harvard, 210: 1–14.

BUTTERFLIES

Since there is a certain amount of West Indian vegetation on the Keys, it is difficult to say whether the exotic Lepidoptera of Cuban or Antillean origin may have become established, or whether they are nothing more than strays of an occasional or perhaps even frequent appearance.

CHARLES P. KIMBALL, 1965

The arthropod phylum includes the class Insecta as its most numerous group. Among the Insecta, the order Lepidoptera has long been the best loved by humans; it includes butterflies and moths. If you are interested in moths, I warn you their numbers are staggering. You might begin with the general references (above) and Mitchell and Zim (below), but soon you will have to repair to *The Moths of America North of Mexico*, which began in 1971 and by 1978 amounted to twenty-two volumes, published by E. W. Classey, Ltd., and the Wedge Entomological Research Foundation, London.

Our most famous butterfly is the Schaus swallowtail, *Papilio aristodemus ponceanus*. It was described and named by Dr. William Schaus of the Smithsonian Institution. He was born in 1858 and wrote 122 papers, describing some 5,000 new insects, before his death in 1942. He described the form we call the Schaus in 1911 from specimens taken at Miami. Today the form is restricted to North Key Largo and Elliott Key. It has been reduced to the edge of extinction by habitat destruction and pesticides. Some people now use the generic name *"Heraclites"* for this and other *Papilio,* but I am informed by my entomological colleagues that this is not a valid genus.

The Schaus is bold yellow and black with bright yellow, scallop-shaped spots along the outside edge of the fore wing. It reaches nearly four inches (9.5 cm) in wingspread. There are related subspecies in the Bahamas, Cuba, and Hispaniola. They differ in coloration and markings. The Cuban form, for example, which I have seen, has broader yellow zones through the fore wing. All of the subspecies of *P. aristodemus* in the Bahamas and Greater Antilles are detailed by Clench, 1978.

Two other swallowtails in the Keys can easily be confused with the Schaus. The giant swallowtail, *Papilio cresphontes cresphontes*, gets much bigger, to five-and-a-half inches (14 cm). The yellow scallops along the edge of the fore wing are reduced to dull smudges. The body is largely yellow. The Bahama swallowtail, *Papilio andraemon bonhotei*, is smaller, to a little over three inches (8 cm) across. It has large, very bright yellow spots along the outer edge of the fore wing and a largely black body.

The Schaus is a creature of tropical hardwood hammock. The larvae feed mostly on torchwood (*Amyris elemifera*), an understory tree that was highly valued for fuel (and torches) early on, and which has been decimated. Unlike its relatives, which frequent open, sunny edges and openings in the forest, the Schaus prefers the deep shade of old growth hammocks. This species' range has already been so curtailed by habitat destruction that I believe any further loss of North Key Largo hammock could easily doom it. Plans calling for "nodes" of development along North Key Largo would not only break up the hammock expanse, but would bring in humans who would clamor for pesticide fogging. I suspect pesticide application would sooner or later span the Key and the fragmented little populations of Schaus would readily succumb. People and old growth hardwood hammocks do not mix well.

The Schaus could easily be re-established in the Lower Keys, for example at Watson Hammock within the National Key Deer Refuge on Big Pine, if only the Fish and Wildlife Service would stop allowing pesticide spraying on its Refuge lands. The situation is ridiculous. The very agency charged with saving endangered species is unwilling to adopt the most obvious reform imaginable to benefit one. Insecticide spraying on National Wildlife Refuges is anathema to sane biologists.

There is no reason why the beautiful Schaus swallowtail need not again be abundant for everyone to see. We need only restore its food plants to large areas of publicly owned hammock and stop spraying chemicals that kill it on its habitat. The money saved in ten years from a cessation of spraying would likely buy all the critical habitat in the Keys.

Many other butterflies live in the Keys, including several not found elsewhere in North America. Their biogeography and evolution seem closely similar to that of birds, which is not surprising. While I know little about most of them, the following works should get you started.

References

Clench, H. K. 1978. *Papilio aristodemus* (Papilionidae) in the Bahamas. *Journal of the Lepidopterists' Society* 32(4): 273–76.

Covell, C. V., and G. W. Rawson. 1973. Project ponceanus: a report on first efforts to survey and preserve the Schaus swallowtail (Papilionidae) in southern Florida. *Journal of the Lepidopterists' Society* 27(3): 206–10.

Forbes, W.T.M. 1941. The Lepidoptera of the Dry Tortugas. *Psyche* 48(4): 147–48.

Kimball, C. P. 1965. *The Lepidoptera of Florida*. Gainesville: Division of Plant Industry, Florida Department of Agriculture.

Klots, A. B. 1951. *A Field Guide to the Butterflies*. Boston: Houghton Mifflin Co.

Mitchell, R. T., and H. S. Zim. 1977. *Butterflies and Moths*. New York: Golden Press.

Pliske, T. E. 1971. Notes on unusual species of Lepidoptera from southern Florida. *Journal of the Lepidopterists' Society* 25 (4): 294.

Rawson, G. W., and W. M. Davidson. 1964. An annotated list of the Lepidoptera observed or collected in 1959–1960 on the Dry Tortugas islands. *Journal of the Lepidopterists' Society* 17(4): 225–27.

Schaus, W. 1911. A new *Papilio* from Florida, and one from Mexico (Lepidoptera). *Entomological News* 22: 438–39.

Tyler, H. A. 1975. *The Swallowtail Butterflies of North America*. Healdsburg, Calif.: Naturegraph Publishers.

DRAGONFLIES

Adult dragonflies are superb fliers and have the potential to disperse over long distances. Thus their mere presence in the salt marsh is no indication of the abilities of the nymphs to survive in seawater. Indeed, most dragonflies cannot complete their life cycles in seawater.

WILLIAM A. DUNSON, 1980

The insect order Odonata shows interesting parallels to the vertebrate order Anura—frogs. All odonates are dragonflies, but some special dragonflies are called damselflies (all toads are frogs). All odonates have an aquatic larval stage, called the nymph. Damselfly nymphs differ from those of other dragonflies in having external gills at the end of their tails instead of internal gills inside the rectum.

All dragonflies are highly predacious, feeding mostly on other insects. They are generally among the most beneficial of man's colleagues on Earth. They are also big, often very colorful, and aeronautic wonders. The aquatic nymphs are predacious too. Like frogs and toads (Anura) the odonates have not been widely successful in colonizing saline waters. There are few marine forms.

In discussing our species, Dunson (1980) points out: "At times adult odonates appear in such numbers on such tiny, isolated islands lacking fresh or brackish water that they apparently represent migrating groups." And ". . .13 species of adult odonates were captured on Summerland Key; only five were collected as nymphs."

Only one of those species, perhaps unique in the world, has nymphs that are truly marine. *Erythrodiplax berenice* nymphs are common in mangroves in full-strength seawater, and have been recorded living in the Dry Tortugas in salinities of 64 ppt—nearly double the salt content of normal seawater.

Here is an annotated list of Dunson's Lower Keys dragonflies:

Anomalagrion hastatum: arrowhead damselfly. Damselflies can be immediately recognized in the field because they hold their wings close to their abdomens (and so fairly close together), and their wings taper approaching their bodies. Their hind wings are not broadly expanded near their bases.

Ischnura ramburii: forktail damselfly. The members of this genus are brightly colored; males with green and blue and rather dark, females paler and with a blue-green wash. Nymphs can live in brackish water.

Brachymesia gravida: Keys red dragonfly. Dragonflies perch with their wings held out at right angles to their elongate bodies (abdomens) and nearly horizontal. The hind wings are broadest near their bases and almost squared off, tapering abruptly along the body to the attachments. I can find no discussion of this species and know only that it occurs in the Lower Keys.

Brachymesia furcata: red-faced dragonfly. This is a large, handsome species, widespread in the Antilles, Bahamas, Mexico, and South America. It is reddish all over. The face is bright red with some yellow trim. The nymphs can tolerate brackish water.

Celithemis eponina: yellow and brown dragonfly. This is a large, abundant species over most of North America. It is colorful, the wings very boldly patterned in yellow and dark brown. There are two brown bars through each of the four wings and a disjunct spot on the median, posterior, broadly expanded part of the hind wing. Pairs are apt to fly together for some time after mating.

Coryphaeschna virens: Antillean green-faced dragonfly. This is a large species, approaching nine centimeters—more than three inches—in length. It flies high and fast. Its close relative, *C. ingens,* is more widespread in the southern United States, and often called "sky pilot." Interestingly, our species, *C. virens,* seems to occur in the United States only in the Florida Keys, where *C. ingens* has yet to be recorded.

Erythrodiplax berenice: marine dragonfly. This is the species with the salt water nymphs studied by Dr. William Dunson. It gets to be about 35 mm long (more than an inch). The freshly metamorphosed adult is boldly striped in yellow and black, but darkens with age. Males become virtually black all over; females retain spots on abdominal segments 3 to 7, which may stay yellow or turn red. This genus is largely tropical, but the species *E. berenice* (maybe the same as *E. naeva*) ranges north to Canada.

Macrodiplax balteata: olive-faced dragonfly. "A fine little Southern species," say Needham and Westfall. It attains about four centimeters (less than two inches) in length, has yellow and black markings, and a red tinge to the front of the thorax. Only two species are in the genus, ours and an Old World form. They are largely tropical, but ours is widespread west to Texas. It also occurs in Cuba, Jamaica, and Hispaniola. The nymphs can tolerate slightly brackish water.

Orthemis ferruginea: black-headed red dragonfly. "A dashing red species, high-perching, strong in action, and swift at dodging," say Needham and Westfall. They go on: "Top of head in mature male becomes wholly metallic black with violet reflections. Thorax and abdomen become red, with a purplish overcast." About a dozen species are in the genus, centered in the tropics. Ours occurs throughout the Greater Antilles and has been recorded not only in Florida but in Mississippi and the Southwest. Nymphs can tolerate brackish water.

Pachydiplax longipennis: white-faced dragonfly. This species and genus are endemic to America, where widespread north to Canada. Adults grow to less than two inches (40 cm). The head is trimmed with cream-yellow and metallic blue. The wings are dark at the bases, often with a yellow tinge. Needham and Westfall note distinctive behaviors of this form. For example: "When two males meet in mock combat, they have the curious habit of facing each other threateningly, then darting upward together into the air and flying skyward, often until lost from view."

Pantala flavescens: cosmopolitan dragonfly. Said to be "found on all continents except Europe" by Needham and Westfall. It also occurs far out at sea. Attaining 50 cm (two inches), this species has confusing facial color; it starts out yellow but becomes red in old males. The body is tawny with black streaks and black crescents over the limb bases. To truly learn dragonfly identification it is necessary to study wing venation: a technical business well described by Needham and Westfall. Despite their almost ubiquitous presence in the most temporary waters, nymphs of this species seem quite intolerant of salt.

Tramea binotata: violet-faced dragonfly. This is a 50 cm (two inch) brown species with a broad brown band across the base of the hind wing. The facial color develops with age, especially in males. There is always a yellow stripe across the face, sometimes a tinge of red on the abdomen. With age, these dragonflies—especially the males—tend to blacken. This species is known from the Greater Antilles (Cuba, Hispaniola, Jamaica, and Puerto Rico) and South America. In North America it is known only from Florida.

Tramea onusta: red-and-green-faced dragonfly. Another 50 cm (two inch) species; largely red all over, including the face—except for the top, which is greenish. The wings are largely brownish with red veins. The broad basal dark band on the hind wing is divided by a clear strip. This species is widespread from Canada to Panama, including at least Cuba and Puerto Rico in the Antilles.

References

Dunson, W. A. 1980. Adaptations of nymphs of a marine dragonfly, *Erythrodiplax berenice*, to wide variations in salinity. *Physiological Zoology* 53(4): 445–52.

Needham, J. G., and M. J. Westfall. 1954. *A Manual of the Dragonflies of North America (Anisoptera).* Berkeley: University of California Press.

BEETLES

"What," the council of distinguished theologians asked, *"can you perceive about the Creator from studying His Creation?"*

The great professor of biology, Dr. J.B.S. Haldane, instantly replied: "An inordinate fondness for beetles."

The order Coleoptera—beetles—is the most prosperous and speciose of all the class Insecta, the most prosperous and speciose of all the classes of phylum Arthropoda—the most prosperous and speciose of all animals on Earth. There are simply lots more beetles than anything else.

Beetles have been loved and loathed by humans throughout our short history, going back five or six thousand years to the sacred scarabs of Egypt, at least. They are apt to be very beautiful, but some are probably our most serious agricultural pests and therefore the most economically important animals alive. Our lives depend on an eternal compromise with beetles. Fortunately, here in the Keys, we do not depend on any significant local agriculture and can enjoy our beetles.

Most insects—like butterflies and dragonflies—have four wings. Beetles differ from other insects in having their forward pair of wings modified into hard shields. These shields close over the usually large, membranous hind wings, which the beetles use for flying. Beetles, like butterflies and dragonflies (but not like roaches or true bugs), have complete metamorphosis: their larvae, called grubs, transform in one huge leap into the adults, which look nothing like their larvae. There are some flightless beetles with reduced flying or hind wings, and tiny shields or fore wings.

Our largest beetle is *Dynastes tityus,* described by Linnaeus himself. It is called the unicorn or rhinoceros beetle because it has huge horns—one on its nose (the smaller) and a huge, double-hooked one on the back that arches out over the head. In the Antillean species I am familiar with, *D. hercules,* only the males have horns. They clamp them on tree branches and fly. This not only makes a loud buzzing sound, but cuts the branch like a lathe. In some cases they spin on a branch long enough to cut it off. I have no evidence that our species performs in this manner.

The unicorn beetle is yellow to pea-green above with blackish dots. It attains a length of two-and-a-half inches (more than 6 cm). Maxwell (cited above) found this species "not very common" and noted only Gainesville and Orlando records. I have found them on Big Pine Key. These wonderful creatures are quite harmless. Their grubs are soil-dwellers and do not attack plants. This beetle belongs to the scarab family, to which I will return shortly.

One of our most dramatically impressive beetles in the Keys belongs to the click beetle family, Elateridae. The grubs are called wire worms. The adults click by snapping the pronotum, which is their middle portion, against the hard sternal part of the abdomen. This not only makes a noise, it flips the beetle through the air. These beetles use this method to right themselves if they fall on their backs. We have a fine, large, near-black species in at least the Lower Keys, with a pale green spot on each side of the pronotum. In the dark these spots glow brightly, giving the beetle the generic name *Pyrophorus*—fire bearer. My entomological colleagues have not been able to give me a specific name for our form, but did give me the reference to the most recent published work: Costa, 1976. This would clearly seem to be an Antillean colonizer in the Keys.

At present, Dr. Stewart B. Peck and colleagues from Carleton University in Ottawa are studying our scarab beetles. Most of these species are dung feeders or scavengers of carrion. Dr. Peck uses many methods to catch them, but a can sunk in the ground and baited with meat and dung seems effective. To quote the abstract of Dr. Peck's most recent paper:

Field work on fifteen islands of the Florida Keys produced thirteen species of scarab beetles. . . . Six of these species represent new records for the Keys. Twenty-three additional species (many of which are synanthropic or tramps), previously recorded from the Keys, were not found. Species-area relationships for the islands form a significant regression line as predicted by equilibrium island biogeography theory. It is concluded that many of the islands have low species numbers either because (1) human habitat disturbance has caused many local species extinctions or (2) species turnover rates (extinction over immigration) are high because of scarcity of suitable hosts or adverse soil conditions. Data from highly disturbed Key West and Stock Island suggest that as species turnover continues higher species saturation levels may be regained through the immigration of synanthropic and tramp species. *This work generally points to the lack of much basic information on scarab beetle bionomics* (italics added).

Synanthropic species are those that live together with humans; I call them edificarian—of edifices. Several phenomena we have become used to in dealing with the natural life of the Keys are evidenced by scarab beetles. For example, there are widely disjunct distributions. The beetles *Copris howdeni* and *Melanocanthon bispinatus* occur in the Lower Keys and on the mainland, like slash pines, deer, and rice rats. Are these disjunct distributions real or the product of incomplete collecting? If some, at least, are real, what accounts for them? There is also Lower Keys endemism: *Ataenius superficialis* is a species to date known from nowhere else on Earth except Big Pine Key. The ringless ringneck snake was in the same situation until its known range was recently expanded to the Torch Keys. Key deer, silver rice rat, and Key rabbit are Lower Keys endemics.

Not to be outdone, the Upper Keys have an endemic beetle also. It is a blind member of family Staphylinidae: *Cubanotyphlus largo*. It is not a scarab. It dwells only in the homes of the endemic Key woodrat, nowhere on Earth except Key Largo. Its relatives are Antillean, like our fruit bat, reef gecko, and Schaus swallowtail.

A final phenomenon of similarity I noted from scanning Peck and Howden's data is a very strong species-elevation relationship. This correlation is better than the species-area relationship, as is the case for some Antillean bats (see account of our fruit bat). In the Keys a few inches of elevation can result in sweeping changes in vegetative associations. A small, high Key like Lignumvitae may have far more species than a huge, low Key like Snipe.

Dr. Peck and his beetles would seem to heap up more evidence for the view that life is worth studying and knowledge is worth accumulating. Patterns emerge. New questions need new answers. Useful generalizations may be formed about the support capacities of ecosystems that will help our own species to survive. Future trends may be predictable and severe problems might be avoided. It is reported that Aldo Leopold, the father of wildlife management, first said: The first principle of intelligent tinkering is to save all the pieces.

References

Costa, C. 1976. Speciation and geographical patterns in *Pyrophorus* Bilberg, 1820 (Coleoptera, Elateridae, Pyrophorini). *Papeis Avulsos de Zoologia,* São Paulo, Brazil, 29: 141–54.

Frank, J. H., and M. C. Thomas. 1984. *Cubanotyphlus largo,* a new species of Leptotyphlinae (Coleoptera: Staphylinidae) from Florida. *Canadian Entomologist* 116: 1411–17.

Peck, S. B., and H. F. Howden. 1985. Biogeography of scavenging scarab beetles in the Florida Keys: Post-Pleistocene land-bridge islands. *Canadian Journal of Zoology* 63(12): 2730–37.

White, R. E. 1983. *A Field Guide to the Beetles.* Boston: Houghton Mifflin Co.

Woodruff, R. E. 1973. *The Scarab Beetles of Florida.* Gainesville: Division of Plant Industry, Florida Department of Agriculture.

SPIDERS AND OTHER ARACHNOIDS

About 30,000 species of spiders have been named so far,
representing what is believed to be about one-fourth of the total.

LEVI, LEVI, AND ZIM, 1968

Leaving behind class Insecta, the six-legged members of the vast phylum Arthropoda, we come to class Arachnoidea—the eight-legged members. They have four pairs of walking legs and often stabbing or pinching members up front, making a fifth pair of obvious appendages. If we were to go on past Arachnoidea we would enter class Crustacea with the ten-legged arthropods like crabs. I have not included those here except in passing, but refer readers to the excellent volume by Voss, cited in the Introduction.

Spiders are extremely abundant, highly conspicuous in many cases, and very beneficial. To quote Benjamin Julian Kaston, who wrote the most detailed account yet produced on American species: "Spiders are exclusively carnivorous, and in fact usually seize only live organisms, as the stimulus that attracts the spider's attention is the movement of the prey. Some of the larger tropical species have been known to catch vertebrates, such as small birds, lizards, and frogs."

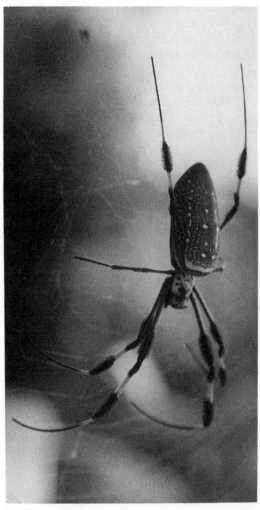

Giant orb-weaver, *Nephila clavipes*. Females may be four centimeters head-body length; males are tiny and hang about the edges of the web. Patterned in lemon-yellow, olive, and gold, these gaudy creatures are fearsome to many people, but are really harmless, highly beneficial insect-eaters of the hardwood hammocks.

Nancy L. Nielsen

Most spiders eat insects. One of ours, the giant hammock spider, *Nephila clavipes*, catches young *Anolis* lizards. Its very close (and no bigger) relative in tropical China catches pipistrelle bats (there is only one dubious record of a pipistrelle in the Keys; see fruit bat account, above). Many people fear spiders but this is a gentle giant, never known to bite humans.

The black widow spider, *Latrodectus mactans*, occurs in the Keys, but is rather scarce. It occurs over a great deal of the world; I have found them from Cape Hatteras to Tasmania. They are jet black, appear quite shiny (rather than furry or hairy), and have red markings on the bulbous abdomen. Black widows build messy webs under things in dark places, like junk piles. The big females (to three-quarters of an inch or 2 cm, body length) may occasionally eat the tiny males, but that habit is by no means standard. Black widows can be deadly, but lethal bites are extremely rare. Levi, Levi, and Zim say, "The patient usually recovers after several days of agony." They go on to point out: "No first aid treatment is available for any spider bite."

As a child growing up part of the time in the deep South, I was always told that the itchy bites I sometimes awoke with in the morning were "spider bites." I am authoritatively informed with no uncertainty that this is incorrect. Spiders do not make itchy bites. Unless you rolled on one that had gotten into the bed quite by accident, you would never encounter one there. Spiders do not bite people if they can avoid it. If a spider does bite you, you will know it: such bites are painful, but very rare.

Our most spectacular and obvious spiders are orb-weavers, family Araneidae. The giant *Nephila clavipes* is our largest example. The much smaller thorny spiders, *Gasteracantha cancriformis* and *Micrathena sagittata* are brightly colored in reds, yellows, black, and white. They spin their webs right across openings—like trails—so as to snag flying insects. *Gasteracantha* is broader than long, with a pair of large thorns (or horns) at each side and a smaller pair behind. *Micrathena* is longer than broad; her thorns, also in three pairs, increase in size posteriorly so that her general shape is like a jagged arrowhead.

The most beautiful and perfect webs are spun by *Argiope* spiders. These produce true orbs: circles of silk attached to spoke-like radiating lines. They make a thick, white, zigzag pattern in the middle of their webs called a signature. Our most common species is the silver orb-weaver, *Argiope argentata*. The females may attain over an inch (3 am) body length and are boldly patterned in black, white, and cream. The males are tiny and glitter like drops of quicksilver at the edges of the female's web.

Our rarest spider, an endangered species, is Irving's cesonia (*Cesonia irvingi*). It is said to be an agile hunter that lives on the ground in sandy soils in leaf litter. Only three specimens were known to G. B. Edwards (in Franz, 1982) when he wrote his account of the species. They had been collected at Key West, Bob Allen Keys, and South Bimini in the Bahamas. This cesonia is pale with a dark stripe down the back interrupted at the rump; a row of dark spots is present along each side. The size (head-body) is less than half an inch: five to nine millimeters.

While spiders have stabbing, fang-like anterior appendages in addition to their four pairs of walking legs, scorpions have pinchers. The pinchers are quite harmless to humans, and scorpions cannot bite us. It is the other end to watch out for: scorpions sting. The last segment at the tip of the tail is modified into a poison capsule with a sharp, curved, injecting stinger. None of our native species is deadly. I have found only two: *Centuroides griseus* is small, about three inches (8 cm) total length (including outstretched tail). Its name *griseus* means grey, but it is amber-yellow to horn color with indications of dark stripes. *C. griseus* is common under litter on the ground, but also climbs well and loves to dwell in hollow branches. It may sting several times in rapid succession. I find the stings less painful than bee stings; they leave me without after-effects in less than an hour. Other people may react more severely.

Our second species, *Centuroides gracilis*, is much larger: up to five inches (13 cm). It is all-over dark brown to nearly black. Its name *gracilis* means slender, but it is much stouter than little *griseus*. I have found *C. gracilis* only on the ground, but it may well climb too. The sting of this species causes prolonged local pain, irritation, and lesions. Victims liken it to being burned with a cigarette.

One of our commonest members of class Arachnoidea is called a grampus or vinegaroon. It looks slightly crab-like in general shape, with long legs. The first pair of walking legs are modified into grand, elongate "whips," which coil like tendrils about stems and fallen twigs in the animal's terrestrial home. Members of family Amblypygidae (which means short-rumped), these creatures possess an anterior, fifth pair of appendages that look at first like fearsome pinchers. They are not, however, and have no ability to inflict damage on humans. Frightening as they look, the amblypygids are harmless. If you uncover one, by all means pick it up and examine it. There is nothing it can do to hurt you, unless you are allergic to mayonnaise. It can secrete a faint odor of this salad dressing.

A very large amblypygid, *Paraphrynus raptator*, attaining more than an inch (29 mm) head-body length, has been collected in Key West. H. V. Weems and C. M. Tibbetts provide an account in Franz (1982). This species is also known from Mexico, and might be merely an introduction in the Keys. However, Weems and Tibbetts note that Florida specimens have shorter spines than the Mexican form.

Another scorpion-like but harmless creature is the giant whiptail, *Mastigoproctus giganteus*. This member of family Mastigoproctidae has false pinchers like an amblypygid, but is more elongate, like a real scorpion. Instead of a segmented tail with a poisonous stinger, however, this animal ends in an antenna-like whip. Without the whip, it reaches about 3 inches; with the whip, twice that and more—to 7 inches (18 cm). Both amblypygids and mastigoproctids are often called "whip scorpions," for obvious reasons. *Mastigoproctus*, however, is the one that can really put out acetic acid when aroused. Acetic acid is what makes vinegar (and mayonnaise) smell, so this animal is most aptly called a vinegaroon.

References

Franz, R. 1982. *Rare and Endangered Biota of Florida, Volume Six, Invertebrates.* Gainesville: University Presses of Florida.

Gertsch, W. J. 1979. *American Spiders*. 2d ed. New York: Van Nostrand Reinhold Co.

Kaston, B. J. 1948. *Spiders of Connecticut*. State of Connecticut Public Document No. 47, Hartford.

Keegan, H.L. 1980. *Scorpions of Medical Importance*. Jackson: University Press of Mississippi. This beautifully illustrated volume is a must for all scorpion aficionados. It includes our *Centuroides gracilis*.

Levi, H., L. Levi, and H. Zim. 1968. *Spiders and Their Kin*. New York: Golden Press.

TREE SNAILS

Many snails are found in trees, but only a few are exclusively arboreal for most or all of their life cycle. Tree snails are normally found on the ground only during egg-deposition or when dislodged from their perches.

JANE E. DEISLER, 1983

Formerly abundant and conspicuous, our tree snails have fared poorly because of their enormous popularity. The shells are often brightly colored and ornately patterned; they are eminently collectible. Snail collectors have wiped out entire populations. Today bulldozers have taken up the project, wiping out entire habitats of hardwood hammock.

These are the only invertebrates I have included in this book that are not arthropods. Snails belong to phylum Mollusca, along with clams, oysters, squids, and octopi. All snails belong to class Gastropoda, which means belly-footed. Those snails that can live in air instead of water are called pulmonates: lung-bearing. Our tree snails belong to family Bulimulidae, a group of neotropical origin that has reached Florida from the Antilles.

One species, *Orthalicus reses*, is federally protected as threatened. The nominate form, *O. r. reses*, is found only on Stock Island. Since it does not live in houses, bars, or warehouses—or on golf courses—you can see just how close to extinction it must be. A few hundred survive on a few dozen trees. Its close relative, *O. r. nesodryas*, differs from the Stock Island tree snail in having a dark apex and dark brown callus. Although *nesodryas* is more widespread, recorded from Lower Matecumbe to Key West, I have never seen it alive in the field. A big *Orthalicus reses* is about 3 inches (7.5 cm) shell length. They are brown and white with flame-like markings.

Stock Island tree snail, *Orthalicus reses reses*.

A second species gets slightly larger. This is *Orthalicus floridensis*. This species is tan with dark brown spiral bands; it lacks the flame-like markings.

The most populous and popular tree snails are genus *Liguus*. There are at present 58 named kinds found in Florida, at least a dozen of which occur in the Keys. Their other major habitats were the Miami rimrock and 'glades hammocks. Much of this original habitat has been destroyed. Many more kinds are found in Cuba. In *Orthalicus*, the shell aperture is half or more of the total shell length. In *Liguus* the aperture is much smaller: less than one-half of shell length. The shell is heavy, like porcelain, and the outside markings do not show clearly through it, as they do in other tree snails. These features, combined with the rainbow colors, make *Liguus* so sought-after.

The classification of these snails has not been worked out well in view of modern notions about genetics and species limits. Most color forms were first described as full species, then lowered to subspecies, and finally pushed into formal taxonomic oblivion as "varieties," like the breeds of dogs. That last step was premature.

Like other tree snails, each individual has male and female parts (*androgynous* is the correct term; "hermaphroditic" is sometimes confused with this condition, but properly refers to sexless intermediates—quite an opposite and freakish phenomenon). However, these snails cannot self-fertilize: they must mate. Therefore, genetic and reproductive criteria are appropriately used in their classification.

Shell collectors have transplanted many of the color forms of *Liguus* to places they never naturally occurred, or into habitats occupied by other members of the genus. Sometimes these transplants were made to save forms about to be exterminated by habitat destruction. At other times a greedy collector just wanted a propagating population closer to home, or transplanted juveniles he hoped would grow to be valuable adults. In any case, the result has been dozens of artificial experiments in genetics and relationships.

Sometimes *Liguus* color forms reside together and never seem to inter-breed, like valid species. In other cases, introgressive hybridization occurs and two color forms are hopelessly merged into a single intermediate type. Occasional hybrids may occur that are new color forms; some of these breed true. Sorting it all out today in terms of species and subspecies will be difficult. Since many forms are now extinct, we may never be able to reconstruct lineages and understand the details of relationships.

For example, I would like to know how many separate invasions of South Florida occurred to give us this diverse tree snail fauna. At what times in the past did these various invasions take place? Perhaps no one will ever know.

References

Clench, W. J. 1946. A catalogue of the genus *Liguus* with a description of a new subgenus. *Occasional Papers on Mollusks,* Museum of Comparative Zoology, 1(10): 117–28, and 1(18): 442–44.

Deisler, J. E. 1983. A key to the tree snails of Florida (Gastropoda: Bulimulidae). *Entomology Circular 246,* Florida Department of Agriculture and Consumer Services, Division of Plant Industry, Gainesville.

Pilsbry, H. A. 1946. *Land Mollusca of North America. Monograph Number 3,* Philadelphia Academy of Natural Sciences, 2(1): 29–102.

OVERVIEW

I now see that the major shift in human evolution is from behaving like an animal struggling to survive to behaving like an animal choosing to evolve. In fact, in order to survive, man has to evolve. And to evolve, we need a new kind of thinking and a new kind of behavior, a new ethic and a new morality.

JONAS SALK, 1984

In fighting nature, man can win every battle except the last.

THOR HEYERDAHL, 1974

This entire book is an extended exposition of two major, integrally related theses: the Florida Keys are utterly unlike the rest of the United States in the richness of their life forms, many of which are found only here. And the Florida Keys represent in microcosm the problems that beset the whole Earth.

The immediate corollaries of those two theses are that as yet we know woefully little about the life forms, and we have been wretchedly inadequate in dealing with the problems.

In building my exposition I have used specific details about animals or their habitats, mentioned wherever appropriate, and thus scattered through the text. It is difficult to pick up these leads and follow them from missing toads to human ethics, but to me the connections are obvious. I must try to make them at least apparent to the widest possible audience. Beginning with notions about the land itself, I see two branches of thought leading alternatively to the way we view biotic diversity and classify the products of organic evolution, and to human ethics and ecology via biogeography. A diagram will help:

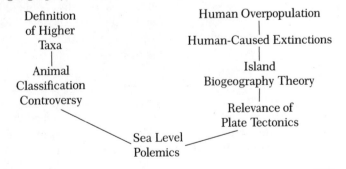

Definition
of Higher
Taxa
|
Animal
Classification
Controversy

Human Overpopulation
|
Human-Caused Extinctions
|
Island
Biogeography Theory
|
Relevance of
Plate Tectonics

Sea Level
Polemics

I can subdivide my theses of uniqueness and syndrome into the seven subjects arrayed in this diagram. With respect to each I have positive views that are often at odds with those expressed even by other biological scientists. I will succinctly rehash those views here, but stress that *in no case* are my opinions novel. I did not invent or codify the positions I hold. Other scientists do agree with me, even if we represent a minority. Literature citations are provided in my text to the back-up works and original ideas I have adopted.

Sea level polemics. The Florida Keys were made underwater. A hundred thousand years ago sea level was about 70 feet (*ca* 21 m) higher than it is today. Land areas worldwide were much smaller than they are today. Therefore, much land occupied by unique animals—like the Keys— simply did not exist. Nor did those unique animals, unless they evolved elsewhere and moved onto new land as it emerged from the sea. From about 60,000 to 12,000 years ago sea level was much lower than it is today. The Florida Keys were high points on a vast peninsula extending into the Antillean–Caribbean Basin. Sea level was more than 300 feet (more than 100 m) below its present level. Land areas were larger and much closer than they are today. Animals could expand their ranges over land, or to islands across relatively narrow water gaps.

I subscribe to the view that sea level has risen continuously since the Wurm glacial maximum, about 12,000 years ago, and today stands higher than it has ever stood since the Sangamon—about 100,000 years ago. No less than Dr. Rhodes Fairbridge disagrees. I see evidence that animals like rice rat and reef gecko colonized the aboriginal Keys as they emerged from the Sangamonian sea, and that many other animals like deer and rabbits remain stranded on the Keys after moving in over land. If Fairbridge is correct, my notions about the times of isolation of the various animals are wrong. Sea level changes, both in the 100,000-year, 400-foot macrocosm, and the 10,000-year, 20-foot microcosm, have been largely ignored by other biogeographers.

Animal classification controversy. Many animals in the Keys are different from their relatives elsewhere. Some are obviously derived from North American ancestors (deer), others from Antillean ancestors (reef gecko). Some are strongly differentiated (rice rat), some show only slight differences (spadefoot); some are therefore called full species, others subspecies, many nothing formal at all: their unusual characteristics are noted, but are inconsistent. Are these differences the results of longer or shorter periods of isolation? Of more or less rapid rates of evolution? Certainly both options are available (especially if one subscribes to my view of sea level rise). To understand the options and make intelligent choices between alternative evolutionary histories, we need the best possible analyses of differences. In some cases it has been questioned whether the differences we see in Keys animals are even genetic, as opposed to environmentally induced (deer, cotton rat). In other cases the differences seem certainly adaptive (raccoon), and in one case we even know a bit about patterns of heredity (greenhouse frog).

We still know little; far too many of our species are critically endangered and may soon become extinct.

Definition of higher taxa. This seems an arcane twig of reasoning to many, but quite a logical end point for me. If we had a solid, consistent framework of animal classification it would at the very least tidy the situation and allow us to get on with the more interesting proximate issues and controversies discussed above. I have insisted on equivocally defined higher taxa that never hybridize (generic issues of bobcat, mangrove snake, and Cuban treefrog). I also insist on following—not abrogating— the rules in the *International Code of Zoological Nomenclature* (generic name of the deer).

Relevance of plate tectonics. I do not doubt that continents and plates have shifted and moved. However, I *know* plate tectonics and continental drift are irrelevant to the distributions of Florida Keys and Bahamian animals. I see no valid evidence that these geological phenomena had anything to do with land animal distribution in the Gulf–Caribbean Basin. I am extremely skeptical that these phenomena had anything to do with the distribution of any vertebrate animals at all: I think they occurred too long ago. Rejecting plate tectonics as a method of achieving animal distri- bution and subscribing to hypotheses of very divergent evolutionary rates lead me to reject popular taxonomic systems like that invented by Willi Hennig and called "cladistics."

Island biogeography theory. Elegant theory espousing abiotic factors of land area, distance between islands (or islands and mainlands), and times of separation integrally depends on clear understanding of sea level changes (and, if you elect to believe in their relevance, plate tectonic movements). I strongly doubt the overall applicability of the theory and find the Florida Keys provide good examples to refute it. These are small islands at the end of a long peninsula, and theoretically should have little species diversity. Instead, they are the most species-rich land areas—per unit area—in the United States. They have been heavily colonized across relatively large, permanent water gaps. Sometimes the theory provides at least a close approximation to reality (beetles), so we cannot discard it entirely. We must do the relevant research to back it up or refute it in particular cases. It serves us as Niels Bohr's model of the atom serves, or the Hardy-Weinberg equilibrium formula for genetics serves: a model to which one compares reality.

Human-caused extinctions. Any broad understanding of life on Earth that helps us in viewing or guiding our own evolution, or maintaining a viable ecosystem, is obfuscated by artificial extinctions. To understand the system, we need all of its parts—or at least as many as are still available.

Artificial extinctions occur much more rapidly than evolution produces new species. That was never true during other periods of natural extinction on Earth because they were spread over millions of years—not a few hundreds or thousands.

A vast wave of extinctions followed the colonization by Amerindians. Far from living in harmony with nature, the first Americans wiped out a wonderful fauna of large animals: sloths, peccaries, camels, and large birds in the area of the Florida Keys. That was probably about three thousand years ago. With the coming of Europeans we lost black bear, monk seal, and such edible birds as Key West quail-dove and zenaida dove. At least the latter two could be restored to the Keys. However, we are in immediate danger of losing lots more of our species: Key deer, Key rabbit, silver rice rat, Key woodrat, Key cotton mouse, Bachman's warbler, Key ringneck snake, Key mud turtle, and Schaus swallowtail—just to list the most immediately obvious. The first Americans wiped out species because they were hungry and had no concept of resource conservation. They ate everything they could kill; their populations boomed and crashed as they colonized, exploited, and devastated each new island or land area. We destroy species for far less compelling reasons: building vacation resorts and retirement homes, and attempting to control insects that irritate us. These activities decrease the ability of the Keys ecosystem to provide us with food. We do not need the ecosystem to feed us today; we import most of our food. I obviously believe that is not a valid reason for destroying the ecosystem or exterminating any of its species. We may again need it.

Human overpopulation. To me this is obviously the root cause of all Earth's problems. Most of our leaders pretend overpopulation does not exist or is a soluble problem. Most seem to think the problem can be solved by putting more land under cultivation, developing more productive plant breeds, and increasing industrial plants. All of these "solutions" merely buy time at the expense of the whole Earth ecosystem. Carbon dioxide builds up as forests are felled and swamps are drained; this produces an obvious greenhouse effect, warming the planet. Sea level rises with aberrant rapidity and desertification expands across much of the land. Oxides of sulfur and nitrogen produce sulfuric and nitric acids, which fall on us as acid rain, as industrial plant and fuel consumption increases. Medical art enables us to breed back into our population all sorts of maladapted genetic constitutions. The resultant problems combine to fuel medical (and other) cost inflation.

The Florida Keys are indeed a mere microcosm of the almost overwhelming problem of overpopulation. Our population in these little isles is exploding, not because of especially high birth rates (though the birth rate is increasing here), but because people move here. Most of the people who move here do not really like the Florida Keys.

It was local biologist George Garrett who first codified this point. The immediate reaction of modern Key colonizers is more brutal than was that of the Amerindians. These are rocky islands, flanked by rocky shores or mangrove swamps, and supporting dense jungles with a bewildering array of tough plants—many poisonous or thorny. Most people loathe the real Florida Keys. They immediately set out to convert the Keys to sandy beaches, lawns and golf courses, and scattered, swaying palms. That would hardly matter if there were only a few people. Our overpopulation problem is habitat destruction, just as surely as if we converted the land to pineapple plantations (which was once begun).

My overview also leads me to make three points that derive from much of what I have rehashed above, but which I believe are novel. I have not heard or read of other scientists specifically deriving these notions. They are disparate, beginning with an evolutionary, anatomical observation about our differentiated mammals and proceeding to views of human activity based on observations of my broadest scope as a world-traveling scientist.

The mammalian fronto-nasal region in South Florida. Our endemic mammals, both in the Keys and on the proximate mainland, tend to have broad or inflated frontals or nasals or both. Examples are Key deer (broad), Key rabbit (inflated), panther (both), and raccoons (both). A mainland example is the round-tail muskrat, *Neofiber alleni*. The silver rice rat of the Lower Keys is an interesting departure: narrow but very elongate nasals.

The lack of rigorous science in conservation efforts. I believe all organizations that claim to exist for wildlife conservation, that claim to be non-profit and therefore tax-exempt (contributions tax-deductible), and/or that spend taxpayers' monies should have to qualify as scientific—*not charitable*—corporations or agencies. I would apply the same standards of peer-reviewed scientific production to government agencies, like the Fish and Wildlife Service and the National Park Service, as apply to private scientific nonprofit corporations. Enforcement of a scientific standard would eliminate—or, better, force radical change upon—many nonprofits that now qualify as charitable. At least two nonprofit corporations active in the Florida Keys have produced real scientific works—and in substantial quantity: The Florida Keys Land Trust and The Conservation Agency. The National Park Service has also produced fine science, but the Fish and Wildlife Service has not lived up to either its potential or its responsibilities. The Nature Conservancy, another private nonprofit, has the potential to accomplish great works in the Keys, but has yet to take a significant role. This is not congruent with its guiding policies and I hope the situation will soon be redressed.

Long-term human survival depends on mimicking natural selection. Natural selection operates in Earth's ecosystems to constantly cull populations and constantly weed out maladapted genetic constitutions. We need both processes desperately. For the short run population reduction would be a huge help; just growth stoppage would go far towards saving the Florida Keys. In the long run, however, we need to plan our evolution in light of genetic knowledge (see, for example, greenhouse frog). In mimicking natural selection we need never employ the simple but brutal methods of nature. We need only persuade lots of people to refrain from reproduction. That works just as well in genetic terms as grisly death itself.

Each of my novel, disparate three points will require far more detailed development than can be provided in an overview of wildlife in the Florida Keys. They will all require more research, but that is the primary exhortation of this book.

References

Heyerdahl, T. 1974. *Fatu-Hiva*. New York: Doubleday and Co.

Oldfield, M. L. 1984. *The Value of Conserving Genetic Resources*. U.S. Department of the Interior, Washington, D.C. "The best overall defense of biological diversity," *fide* B. Hough-Evans, 1987, *American Scientist* 75(2): 204–5.

Salk, J. 1984. Dr. Jonas Salk's formula for the future. *Parade*, November 4: 9–10.

Umpleby, S. A. 1987. World population: Still ahead of schedule. *Science* 237(4822): 1555–56.

INDEX

Note: Page numbers in italics refer to illustrations.

About the Author

James Draper Lazell, Jr., was born on Manhattan Island in 1939. His family moved to Philadelphia, Pennsylvania, where he attended Germantown Friends School, graduating in 1957. His bachelor's degree is in English from the University of the South, Sewanee, Tennessee. He received his master of arts from Harvard, his master of science from the University of Illinois, and his Ph.D. in biology from the University of Rhode Island. Since early childhood he has been possessed by a passion for animals. The day he graduated high school he left on an expedition to the West Indies and, he says, "in many ways I never came back."

Lazell has taught at the University of Massachusetts and Tufts University, but does research full-time now. He is an officer of Harvard University and an associate of the Museum of Comparative Zoology there. In addition, he is curatorial affiliate for recent vertebrates at the Peabody Museum of Natural History at Yale University, on the staff of the Bishop Museum, Honolulu, Hawaii, and also Life Fellow of the Explorers Club.

Lazell has published over a hundred articles and technical reports. His first book, *This Broken Archipelago*, about the reptiles and amphibians of Cape Cod and the Massachusetts coastal islands, was published in 1976. Lazell's researches are now concentrated on the West Indies (including the Florida Keys), various Pacific islands, Australia, and tropical Asia. His legal residence is on Conanicut Island, off the coast of Rhode Island, but he is more often far away, in a small boat on a remote sea, heading for some other island, all too little-known.

ALSO AVAILABLE
FROM ISLAND PRESS

Americans Outdoors: The Report of the President's Commission
The Legacy, The Challenge, with case studies
Foreword by William K. Reilly
1987, 426 pp., Appendixes, case studies, charts
Paper: $24.95   ISBN 0-933280-36-X

The Challenge of Global Warming
Edited by Dean Edwin Abrahamson
Foreword by Senator Timothy E. Wirth
1989, 350 pp., tables, graphs, index, bibliography
Cloth: $34.95   ISBN: 0-933280-87-4
Paper: $19.95   ISBN: 0-933280-86-6

A Complete Guide to Environmental Careers
By the CEIP Fund, Inc.
1989, 275 pp., photographs, index, references
Cloth: $24.95   ISBN: 0-933280-85-8
Paper: $14.95   ISBN: 0-933280-84-X

Crossroads: Environmental Priorities for the Future
Edited by Peter Borrelli
1988, 352 pp., index
Cloth: $29.95   ISBN: 0-933280-68-8
Paper: $17.95   ISBN: 0-933280-67-X

Down by the River: The Impact of Federal Water Projects and
Policies on Biodiversity
By Constance E. Hunt with Verne Huser
In cooperation with The National Wildlife Federation
1988, 256 pp., illustrations, glossary, index, bibliography
Cloth: $34.95   ISBN: 0-933280-48-3
Paper: $22.95   ISBN: 0-933280-47-5

From THE LAND
Articles compiled from THE LAND, 1941–1954
Edited and compiled by Nancy P. Pittman
New Introduction by Wes Jackson, The Land Institute
Conservation Classic Edition 1988
459 pp., line drawings, index
Cloth: $34.95   ISBN: 0-933280-66-1
Paper: $19.95   ISBN: 0-933280-65-3

Holistic Resource Management
By Allan Savory
Center for Holistic Resource Management
1988, 512 pp., plates, diagrams, references, notes, index
Cloth: $39.95   ISBN: 0-933280-62-9
Paper: $24.95   ISBN: 0-933280-61-0

Last Stand of the Red Spruce
By Robert A. Mello
Introduction by Senator Patrick J. Leahy
In cooperation with the Natural Resources Defense Council
1987, 208 pp.
Paper: $14.95   ISBN: 0-933280-37-8

Our Common Lands: Defending the National Parks
Edited by David J. Simon
Foreword by Joseph L. Sax
In cooperation with the National Parks & Conservation Association
1988, 575 pp., index, bibliography, appendices
Cloth: $45.00   ISBN: 0-933280-58-2
Paper: $24.95   ISBN: 0-933280-57-2

Pocket Flora of the Redwood Forest
By Dr. Rudolf W. Becking
1982, 237 pp., drawings, photographs, index
Paper: $15.95   ISBN: 0-933280-02-5

Saving the Tropical Forests
By Judith Gradwohl and Russell Greenberg
Preface by Michael H. Robinson
Smithsonian Institution
1988, 207 pp., index, tables, illustrations, notes, bibliography
Cloth: $24.95   ISBN: 0-933280-81-5

Sierra Nevada: A Mountain Journey
By Tim Palmer
1988, 352 pp., illustrations, appendices, index
Cloth: $31.95   ISBN: 0-933280-54-8
Paper: $14.95   ISBN: 0-933280-53-X

For a complete catalog of Island Press publications, please write:

Island Press
Box 7
Covelo, CA 95428